Management for Professionals

More information about this series at http://www.springer.com/series/10101

Gunther Friedl • Horst J. Kayser
Editors

Valuing Corporate Innovation

Strategies, Tools, and Best Practice From the Energy and Technology Sector

Springer

Editors
Gunther Friedl
TUM School of Management
Technical University Munich
Munich, Germany

Horst J. Kayser
Siemens AG
Munich, Germany

ISSN 2192-8096 ISSN 2192-810X (electronic)
Management for Professionals
ISBN 978-3-319-64863-7 ISBN 978-3-319-64864-4 (eBook)
DOI 10.1007/978-3-319-64864-4

Library of Congress Control Number: 2017951446

Printed on acid-free paper

This Springer imprint is published by Springer Nature
The registered company is Springer International Publishing AG
The registered company address is: Gewerbestrasse 11, 6330 Cham, Switzerland

Preface

Innovation is one of the key economic factors for society's progress. Technology-based corporations are important drivers of innovation. Global competition forces them to be constantly successful with their innovative activities. Companies that fail in their innovative efforts will be pushed out of the market, frequently, when others come up with better solutions. To successfully and sustainably innovate, companies need concepts and tools for assessing their innovative activities.

Even though academic research provides such concepts and tools, many of them are not in use in corporate reality. Corporations frequently use their own concepts and tools. However, they usually do not report about their successful concepts.

Valuing corporate innovation is of utmost importance for any technology-based cooperation. We aim to bridge the gap between theory and practice in this area. We not only discuss current academic research in selected fields but also present case studies carried out at Siemens; these provide detailed insights into approaches and tools that are applied in practice. In doing this, we hope to foster closer cooperation between academia and the corporate world. Each can learn from the other, and we believe that this learning experience is inspiring and valuable for both sides.

We want to express our sincere thanks to everybody who contributed to this book: the authors of the three parts, Dr. Peter Schäfer, Dr. Christopher Scheubel, and Friedrich Walcher, researchers at Technical University of Munich, as well as Philipp Bierschneider, Holger Gierse, Dr. Ralf Hermann, Ulrich Wöhrl, and Michael Wokusch, experts at Siemens AG.

Our further thanks go to Willibald Fischer, Stefan Jung, Sonja Neidhardt, Dr. Hans Rauner, Prof. Dr. Klaus Riedle, Prof. Dr. Jörg Sauerbrey, and Stefan Schardt, all of whom supported us greatly in developing this book.

Munich, Germany — Gunther Friedl
Munich, Germany — Horst J. Kayser
March 2017

Contents

About the Editors

Gunther Friedl has been dean of the TUM School of Management at Technical University of Munich since 2010. He joined TUM as a full professor of management accounting in 2007. Gunther Friedl holds both a Diploma (M.Sc.) in business administration from the University of Munich and a Diploma (M.Sc.) in physics from Technical University of Munich. He received his Dr. (Ph.D.) and Habilitation in business administration from the University of Munich.

Before joining the Technical University of Munich, Gunther Friedl has been full professor at the University of Mainz. He has been visiting scholar at Stanford University and guest professor at Warsaw School of Economics.

Gunther Friedl is author and co-author of several textbooks and monographs on management accounting. His current research interests lie in the area of executive compensation, performance measurement, and valuation. His work has been published in journals such as European Journal of Operational Research, OR Spectrum, Research Policy, and Schmalenbach Business Review.

Horst Kayser has been Chief Strategy Officer of Siemens AG since 2013. He studied electrical engineering and business administration at Technical University of Darmstadt. Horst Kayser gained his Master of Public Administration at Harvard University and obtained a Ph.D. in business administration at the University of Mannheim.

After having worked for McKinsey & Company for six years, he joined Siemens Management Consulting (SMC) in 1995. He became managing partner of SMC in 1998. Afterward, he held several positions at Siemens AG. Amongst others he was Chief Executive Officer of Siemens Ltd. South Korea and Chief Executive Siemens UK. Before his current position as Chief Strategy Officer of Siemens AG, he was Chief Executive Officer of AEG Power Solutions B.V. and of KUKA AG.

Complementary to his work on the board of Siemens AG, Horst Kayser is member of the supervisory board of KENDRION N.V.

Introduction

Gunther Friedl and Horst J. Kayser

Corporate innovation and corporate entrepreneurship subsume all organizational activities ensuring sustained company success. They include product development, product improvement and process refinement, all of which require considerable amounts of management commitment and company resources (see Schollhammer 1982, p. 211). Innovation in product development and improvement is one of the key factors determining the success of any corporation. The ability to innovate is of particular importance for technology based corporations who constantly need to reinvent their product portfolio to keep pace with changing customer preferences. At the same time, they must continuously innovate their processes to ensure their cost competitiveness.

While modern society tends, not without reason, to emphasize the role of individual entrepreneurs in innovation, the role of corporate entrepreneurs and corporate innovation is less well recognized. However, the important role of corporate entrepreneurship in terms of innovation is well documented in the economics literature. As early as 1942, in his seminal work "Capitalism, Socialism and Democracy", Joseph Schumpeter pointed out that large companies are the most important drivers of economic development and innovation. These companies have the necessary resources to invest in the research and development needed to create better and less expensive products—a fact that is still valid today. In Germany, for example, large corporations with more than 500 employees account for 60% of the revenues and 39% of the employees of all German companies (see Günterberg 2012, p. 3). At the same time, they are responsible for 84% of all internal corporate research and development expenditures, which amounted to EUR 51.1 billion in 2011. Moreover, they account for 90% of all external research and development

G. Friedl (✉)
TUM School of Management, TU München, München, Germany
e-mail: gunther.friedl@tum.de

H.J. Kayser
Siemens AG, München, Germany

G. Friedl, H.J. Kayser (eds.), *Valuing Corporate Innovation*, Management for Professionals, DOI 10.1007/978-3-319-64864-4_1

expenditures, i.e. funding of third-party research and development of German companies (see Stifterverband 2013, p. 9). Not surprisingly, economic and management research has devoted a lot of attention to corporate innovation and corporate entrepreneurship.

Since corporate innovation consumes vast resources, an important research question is how to best allocate these resources among different innovation projects. In order to allocate them as efficiently as possible, companies need valuation tools to assess the benefits and costs of these projects. This is particularly important in the case of technological innovations, as they usually require very high investment.

Corporate innovations are subject to different kinds of uncertainties. Therefore, these factors must be included in any project valuation tool. Different types of uncertainty require different methods. Technological uncertainties are highly company-specific. Estimating these uncertainties requires expert opinions and is usually quite subjective. Market uncertainties can be assessed by observing market prices, which are a more objective measure. Political uncertainties usually depend on the political system, the stability of the current government, and the economic situation of the country.

These uncertainties are fundamentally determined by major societal challenges, including climate change, population growth, and the digitalization of society. The accurate measurement of uncertainties can be improved by differentiating between technological, market, and political uncertainties. The energy sector is an illustrative example of this structured approach. Global climate change and population growth drive the need for new technological solutions with high technological risks. Simultaneously, there is a lot of market information, such as oil and gas prices or the day-ahead market for electricity, which allow to obtain objective data on market uncertainties. Legislative reforms such as the German "Erneuerbare-Energien-Gesetz (EEG)" accompany technological changes, and greatly influence the profitability of research projects.

Therefore, valuation tools for research and development in energy markets must incorporate all sources of information to wholly assess the overall uncertainty. The objective of the first part of this work (Schäfer) is thus to identify important components of a valuation tool that is able to incorporate technological as well as market and political uncertainty in order to improve resource allocation for projects in research and development. The example of the Siemens H-class gas turbine is used to demonstrate the valuation process for a long-term innovation project in the area of energy markets. The case study shows that theoretical considerations can be fruitfully applied in a practical setting.

Another important innovative sector worth investigating in terms of valuation, is product lifecycle management. Digitalization affects all elements of the value chain, such as supplier integration, manufacturing, sales, administration, as well as research and development. Successful innovation processes require the integration of all major stakeholder groups such as developers and engineers, product managers, suppliers and customers. Cloud-based collaboration solutions can significantly improve innovation processes and is therefore a main source of value creation. Product lifecycle management has the potential to change the competitive landscape of innovation. Despite huge upfront investments, product lifecycle management investments will provide sustainable competitive advantages. To describe the objectives, processes and the potential of product lifecycle management is the main objective of the second part of this work (Scheubel, Bierschneider, Gierse, Hermann, Wokusch). Again, a Siemens case study is used to demonstrate the opportunities in this recent development.

While both topics—the valuation of product development projects under uncertainty and product lifecycle management—are important on a project basis, entire companies need a comprehensive assessment of their innovation power. Measuring the performance of innovative activities is of crucial importance for technology firms with major research and development projects. The larger the organization, the more difficult it is to assess which areas are the most innovative ones. An important task is to know whether innovation activities are efficient in the sense that creative efforts are concentrated on the most promising projects. The objective of the third part of this work (Walcher, Wöhrl) is to provide a comprehensive measure of innovation performance. They also use the case of Siemens to demonstrate how the innovation power of a global technology firm can be measured and controlled.

The research aims to develop concepts and tools for valuing and controlling innovative activities. One of the main focuses of this work is on the practical applicability of the developed concepts. Therefore, findings from the existing literature are combined with several expert interviews in order to assure the integration of practical requirements in the research and development process. Three case studies illustrate the existing practice in valuing and controlling corporate innovation, and allow to demonstrate that there is a strong link between theoretical concepts of innovation and their practical application.

References

Günterberg, B. (2012). *Unternehmensgrößenstatistik: Unternehmen, Umsatz und sozialversicherungspflichtig Beschäftigte 2004 bis 2009 in Deutschland, Ergebnisse des Unternehmensregisters (URS 95)* (No. 2). Daten und Fakten, Institut für Mittelstandsforschung (IfM) Bonn.

Schollhammer, H. (1982). Internal corporate entrepreneurship. *Encyclopedia of Entrepreneurship, 209*, 223.

Stifterverband für die deutsche Wirtschaft. (2013). *FuE-Datenreport 2013. Analysen und Vergleiche.*

Valuing Research and Development Projects in Energy Markets

Peter Schäfer

1 Characteristics of Research and Development Projects

In this chapter, we highlight the importance of appropriately valuating research and development investments. We analyze the most important characteristics of projects in research and development. Projects in research and development are usually highly uncertain. They have a high degree of managerial flexibility, but investment expenditures are less reversible than e.g. capital expenditures on property, plants or equipment. We will accordingly analyze the typical types of uncertainty in research and development projects and suggest different systematizations of types and sources of uncertainty. Further, we highlight the importance of flexibility in such processes. We close the chapter by examining different types of flexibility, such as abandoning the project after certain important steps when new information emerges.

1.1 The Importance of Research and Development Projects

The development of new products or production technologies is of crucial importance for companies. Research and development activities ensure the competitiveness of a company and thereby play a vital strategic role in the company's success. Particularly in fast developing industries, a large portion of sales are generated by relatively new products. Being first to put an innovative product on the market is one of the most successful sources of competitive advantage. However, corporate innovation and activities in research and development consume huge amounts of resources. They are related to high capital spending. Since companies have only

P. Schäfer (✉)
TUM School of Management, TU München, München, Germany
e-mail: peter.schaefer@tum.de

G. Friedl, H.J. Kayser (eds.), *Valuing Corporate Innovation*, Management for Professionals, DOI 10.1007/978-3-319-64864-4_2

limited capital resources, a reliable valuation of research and development projects is of utmost importance for all companies. The main objective of this allocation process of limited resources to the development projects is to increase the firm's entire value. Appropriate valuation tools ensure optimal investment decisions in research and development projects, and are thereby the key for an efficient allocation of the resources that a company spends on the development of innovative products. Additionally, due to the frequently long development phases and the huge amounts of necessary resources, well-developed controlling and project management is the backbone of a successful research and development process and the company's resulting success.

Projects in research and development have special characteristics that make their valuation and controlling exceptionally difficult. In particular, standard valuation tools, such as net present value techniques, fail to capture all aspects of such projects. The inherent characteristics of development processes are several substantial uncertainties. These can, for example, include the level of capital expenditure that is necessary for the whole development process, the technological success of the development efforts, or the market success of the innovative product. An important driver of uncertainty is the time lag between the development decision and the marketability of the developed products. Depending on the type of product development activity, this time lag can be quite long. Uncertain future price developments of necessary raw materials or of potential substitutes for the developed product are important drivers of project value. These uncertainties present a special challenge for the valuation of research and development projects.

Another characteristic of development projects that a valuation tool must be able to capture is the set of potential actions and flexibility that management have during the development process. While there is a high degree of ex-ante uncertainty, management can react to emerging information during the development process. There is a large range of examples of such flexibilities and potential actions. For example, a substantial change in the structure of material costs can make an alternative technical solution more attractive. Further, the developed product can be more attractive for other markets than the one it had been planned for originally. Farthest reaching, technological difficulties or market changes can make it necessary to stop the development at certain milestones. Obviously, a valuation tool must consider such flexibilities.

The relevance of uncertainty and flexibility is increased by the usually long time lag between the initial investment expenditure and the first cash inflows from the project. Moreover, even the actual development period is uncertain and can be influenced by higher or lower investment expenditures. A further characteristic of research and development projects is the dynamic during the development phase. Effective accomplishment of milestones and monitoring the development of relevant risk factors during the development phase is key for the success of the project (see Granig 2007, p. 52).

1.2 Identification and Classification of Different Sources of Risk in Development Processes

As pointed out in the previous section, incorporating the uncertainties and risks of a research and development project in its valuation is crucial. Decisions about the realization of research and development projects have to consider many different sources and types of risk. For example, it may not be clear how the market for a new product will develop, how competitors will react to the company's decisions, what government and regulating agencies do, or what volume of resources is necessary to successfully develop the new product. Due to the usually high and typically irreversible investment expenditures, a failure to include a proper assessment of such risks in the decision-making process can become a substantial threat to the whole company. Thereby, it is important to identify the relevant risks, to assess these risks and finally, to manage and mitigate the relevant risks throughout the whole project.

When valuating investment projects, a company typically compares the expenditures that are necessary for the project with the cash inflows resulting from the realized project. However, neither the expenditures nor the actual cash inflows are usually known before the project is realized. Therefore, in the first place, one can distinguish risks that affect the expenditures, or cash outflows, and risks that affect the cash inflows. An exemplary systematization of types of uncertainty is given in the Fig. 1 below.

There can be different sources of risk in cash inflows. When the company has to make the investment decision for a potential development process it usually knows neither the quantities of the developed product that will be sold nor the prices that can be obtained for the product on the market. The price-demand curve can provide an idea of the interdependency of price and the quantity sold. However, even the course of the price-demand function will usually be uncertain at the early stage when the investment decision has to be made. For example, the price-demand curve for technical investment goods will depend on the development of the market for

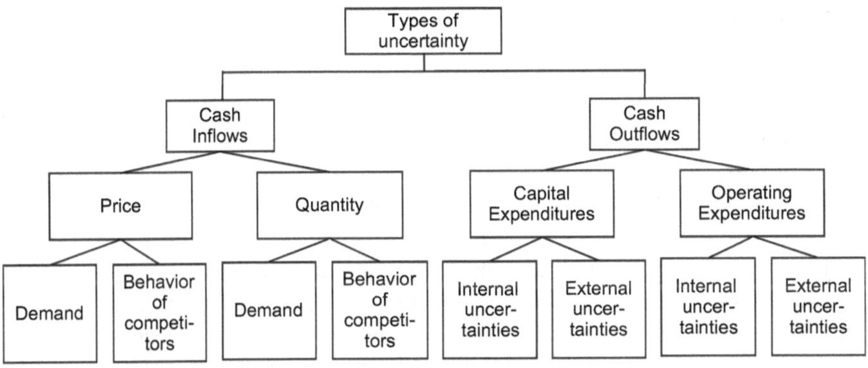

Fig. 1 Systematization of types of uncertainties in research and development projects (Source: Friedl 2001)

the goods that are produced therewith. Other uncertain determinants of the demand for the products can be the clients' preferences as well as the development of the whole economy (see Friedl 2001, p. 27). It is also uncertain how competitors will react to the development of new products. The behavior of the innovator's competitors can influence both the quantity sold and the price of the newly developed product.

Regarding expenditures for the project, one can distinguish capital expenditures and operating expenditures. Uncertainties in both types of expenditures can stem from internal and external sources. External sources could be costs of inputs necessary for the production of the newly developed product. Internal uncertainty in the operating expenditures can include the uncertainty about the quantity of production factors that are necessary to produce the product. But even capital investments can be uncertain. In particular, it is often unclear how long the development process of a certain product will last, and which resources the development process needs.

For an analysis of the risk factors in the course of the valuation, it can be helpful to distinguish different types of risk. In research and development projects there are typically market-related risks. They contain all kinds of uncertainty that relates to the question of what the company can earn with the product during its life cycle. Further, there are different technological risks. This means that certain properties of the planned product cannot be realized or that its development becomes more complex and time-consuming than expected. Besides the market-related and technological risks in research and development projects there can be regulatory risks. During the long development periods the regulatory environment can change and new regulations can influence the marketability and success of the product.

For the valuation of uncertainty, it is important to establish whether a market price for the uncertainty exists or not (see Friedl 2001). By definition, in a complete capital market, a market price for each uncertainty exists. If such a market exists for a particular risk, the risk is referred to as a market risk. An obvious example of a market risk is the purchase of a company's stock. The future cash inflows from the company are uncertain, but the current stock price reflects these risks. However, in reality it is hardly conceivable that each uncertainty can be valued by market prices. For example, consider a company that develops a completely new product. There is substantial uncertainty as to whether the development will fail or can be successfully completed. Moreover, the time to completion is uncertain. These kinds of technological risks are typically private risks for which no market exists (see Smith and Nau 1995, p. 807 or Amram and Kulatilaka 1999, p. 56). Usually there is no information from the market such as a stock price that can be used to assess this risk.

The existence of market prices enables a quantitative provision for market risks in valuations of research and development projects. Consider for example the need for a certain amount of a raw material for the production of the developed product. For example, the price of a call option for the respective raw material that can be derived by arbitrage arguments from the price development of the underlying raw material gives a market price incorporating the risk of a significant change in the price for the respective raw material. In contrast, it is more difficult to take private risks into account quantitatively. Typically, the evaluation of private risks, such as technological risks, depends on the estimations of the decision-makers. Estimations of experts or experiences gained from earlier projects can help to evaluate probabilities for certain private risks, such as the technical feasibility of a planned product. Another way to estimate private risks is to rely on historical information. For example, a company can analyze how often the development of new products in a certain branch had been successful in the past. The share of successful development projects in the past may be a useful approximation for the probability of success of the current project.

To sum up, risk and uncertainty are unavoidable characteristics of innovative research and development projects. A company should carefully identify all relevant risks and consider them in the investment decision. Later in the development process, consistent risk management can reduce the risks or the negative effects of a potential occurrence of a risk.

1.3 Multiple Stages and Flexibility in Research and Development Projects

In the previous section we identified uncertainty as a major characteristic of investments in research and development projects. While cash inflows as well as cash outflows from such projects are usually uncertain, the company often has the flexibility to react to events and information appearing during the development process. In particular, the assumption of a single initial investment that is often made in classical capital budgeting and investment theory is overly simple for this kind of investment projects (see Friedl 2001, p. 19). Instead of a single initial investment, capital expenditures actually occur in multiple stages during the investment period. This allows for an obvious kind of flexibility: For each investment stage the company can usually decide whether it wants to continue the project and spend further money or to abandon the project.

In addition to the drastic measure of abandoning the whole project, the company often has other flexibilities with which to react to new information emerging during the development process. For instance, it might be valuable to delay the further development until new information has been gathered. A particular change in the regulatory environment might be expected which would influence the economical profitability of the developed product. In such a situation, a company can defer the further capital expenditures until clarity regarding the legal regulations has been achieved.

We have already discussed the fact that it is not or cannot necessarily be specified in advance on which exact market or market segment a developed product will be offered. During the development process, it can turn out that an innovative technology can be used for different new products, or have a higher economic impact in another product or market than originally planned. Greater technological success or changes in the target market, such as an expected increase in customer demand can make an expansion of the project or an acceleration of the development advisable. These possible actions and alternative decisions facing a company during the development and investment process are often referred to as real options. The different kinds of real options and potential approaches to value the advantage that a company achieves from the flexibilities will be analyzed more systematically in Sect. 2.3.

Besides the described potential reactions concerning market and investment decisions, there are usually many technological details that are not, or not precisely specified when a project is first considered. So far we have discussed business, investment and market decisions that have to be made and allow for a certain degree of flexibility during the development phase. Besides these, there are technological degrees of freedom that allow for some flexibility depending on new information emerging during the development process. Most importantly, such information can affect development success. If it turns out that a certain technological specification is not obtainable by the planned technical realization of the product, it may be necessary to find new possibilities for the technical realization or to change the intended technological specification of the product. Furthermore, the company may recognize previously unconsidered future needs of the market that makes it worth changing the technological specification of the developed product. When making the initial investment decision, the company must be aware of the economic and technological flexibilities and their interactions in order to make an optimal investment decision.

The degree of reversibility of investment expenditures is another important determinant of the value of research and development projects. Usually investments in property, plant and equipment have a relatively high degree of reversibility, because the respective goods are marketable and can be sold by the investing company. The more specific an investment is for a certain industry or even company the more likely it is to be irreversible (see Dixit and Pindyck 1994, p. 8). Investments in research and development in particular are often not reversible. The most obvious example is the investment expenditure in a research and development project that turns out to be unsuccessful. In addition, unfinished development activities are of course not marketable for the company. Thus, investments in a development process that has already been carried out are hard to reverse. On the other hand, the multiple stages structure of such investments projects allows for a postponement of expenditures. It is not necessary to make the full expenditures at the beginning of the development process. Therefore, irreversibility can be reduced by staggering investments over several phases.

2 Tools for Valuing and Managing Investment Projects in Research and Development

In the previous sections we identified and explained the most important characteristics of investment projects in research and development. Frequently, the expenditures in research and development projects are to a high degree irreversible. Moreover, project expenditures and project revenues are risky. Furthermore, the sequential nature of investment in innovation and development projects allows management diverse economic and technological flexibilities during the development process. This section describes the shortcomings of prominent tools for capital budgeting decisions in research and development. Moreover, we present modern techniques for such valuation tasks, and in particular propose real option approaches as a technique for improving research and development valuations.

2.1 Shortcomings of Classical Valuation Tools for Projects in Research and Development

One of the most important tools for capital budgeting is net present value analysis. The net present value offsets current and future expenditures against forecasted future cash inflows generated by the respective project. The net present value analysis incorporates the project risk in a very simple way. Future cash flows are discounted using a project-specific discount rate that accounts for the riskiness of the project. The higher the risk is, the higher is the discount rate, and the lower is the present value of future cash flows. Indeed, the net present value approach is superior to alternative valuation measures under the assumption of a single stage investment without any flexibility (see Trigeorgis 1996, p. 24). However, this approach has several shortcomings when applied to a project in an uncertain environment with managerial flexibility after the initial investment decision.

Firstly and most importantly, this is because the net present value analysis is a static tool. Only one major project decision based on the information available at project outset is included in the analysis. Thus, the net present value analysis cannot incorporate future managerial flexibility during the course of the project. Hence, several authors argue that net present value approaches might lead to a systematic undervaluation of innovative research and development projects with a high degree of managerial flexibility during the development phase. Suppose, for example, the simple option of abandoning a development project at a certain milestone. The company would exercise this option and abandon the project if the recalculated net present value of the project at this milestone were negative. As a consequence, future cash inflows and cash outflows after the abandonment decision would be zero or at least significantly reduced. Since the net present value analysis cannot integrate this flexibility, it is not clear whether a project with a negative net present value is really a bad one.

Secondly, it is very difficult to capture different types of risk in net present value approaches. Usually expected future cash flows are discounted with a risk-adjusted discount rate. Market risks are comparably easy to capture in a risk-adjusted discount rate. For example, the Capital Asset Pricing Model (CAPM) offers a method for calculating a risk-adjusted discount rate by taking into account the correlation of a certain asset with the market portfolio. The correlation of a risky asset with the market portfolio is the main factor for the risk-adjusted discount rate in the Capital Asset Pricing Model. Hence, a condition for this approach is that a marketable security for the respective risk exists. Otherwise the Capital Asset Pricing Model would not be able to capture the respective risks because the correlation with the market portfolio cannot be calculated.

Suppose, for example, the technological risk that the development of an innovative product will fail. If the development fails, none of the forecasted cash inflows from the project can be realized. However, there is no marketable security that reflects this risk and it is hard to appropriately adjust the discount rate for it. In general, market risks are easier to consider by risk-adjusted discount rates whereas technological or political risks are hard to capture by discount rates. To sum up, the net present value approach is a tool which is superior to other valuation tools in an environment with no managerial flexibility after the initial investment decision, and the presence of mainly market risks that can easily be accounted for by risk-adjusted discount rates.

Alternative approaches attempt to better incorporate uncertainties and the influence of managerial flexibility on the value of investment projects. A frequently used tool is sensitivity analysis. This tool analyzes the influence of a change in a primary variable on the value of the project (Trigeorgis 1996, p. 52). This approach helps to identify the variables that can lead to a substantial threat to the success of the project. An example of a main value driver of a development project is the price level of the innovative product. A sensitivity analysis shows how the value of the project changes when the assumptions for the price level change. An important factor for the decision may be the critical price level that is necessary to achieve in order to make the project profitable. The sensitivity analysis helps to better understand the influence of certain primary variables on the project's value. However, it still has considerable shortcomings. For example, it fails to capture interdependencies between different uncertain variables that commonly influence the value of the project. Further, the scenario analysis is hardly able to capture other qualitative risk factors such as the development's success. Most important, it is still a static tool, which is not able to take managerial flexibility during the development process into account.

The lack of recognized interdependencies and technological risks can be addressed by scenario analysis. In contrast to sensitivity analysis, which analyzes the isolated influence of a single primary variable, scenario analysis incorporates uncertainty by comparing the project's value in different scenarios. These scenarios can differ in all primary variables and assumptions. However, each scenario still considers a fixed investment plan with no managerial flexibility. Scenario analysis does not allow for an explicit consideration of managerial flexibility during the investment period (see Amram and Kulatilaka 1999, p. 39). Scenario analysis can be extended by a simulation analysis. This allows analyses of a very large number of possible scenarios for the uncertain variables. However, this does not solve the problem of allowing for managerial flexibility and decision opportunities after the initial investment decision. In the next section, we discuss two tools that can explicitly take into account these flexibilities.

2.2 Capturing Risk and Flexibility with Decision Tree Analysis

An alternative tool for capital budgeting in research and development projects is decision tree analysis. It makes a first step towards fully recognizing managerial flexibility in research and development projects. This approach addresses the two main shortcomings of net present value analysis. Firstly, in a decision tree analysis it is possible to capture and differentiate different types of risk. Secondly, one can explicitly take managerial flexibility into account in a decision tree analysis. It allows inclusion of potential subsequent decisions and their effects on the project's value (see Trigeorgis 1996, p. 57).

In order to use decision tree analysis, the decision situation must be carefully modeled. A model of the decision situation includes three components: decision nodes, event nodes, and terminal nodes. A decision tree always starts with a decision node, which represents the first decision. For example, the initial decision of whether or not to invest in a new product development project is one node with two possible decisions—investing or not investing. Subsequent decisions, such as abandoning the project, are also modeled as decision nodes. Uncertainty is modeled using event nodes. They represent events that the decision-maker cannot control. Whenever uncertainty resolves, you need to model this with an event node. For example, whether a certain technological specification can be achieved or not is an event node with two possible outcomes. Decision nodes are usually preceded by event nodes because uncertainty makes the option to decide valuable. Finally, the payoff is modeled at the end of each branch in the decision tree. This payoff is modeled as a terminal node. For example, a terminal node can represent the value of a successful product development, taking into account all future revenues and operating expenses.

In a decision tree, the value of the project and the optimal actions contingent on the states of nature are determined simultaneously and backwards. For each decision node, the value-maximizing action is determined. There are two basic ways to value uncertain future cash flows at an event node of a decision tree. The first possibility values uncertain future cash flows with their certainty equivalents and discounts them at the risk-free discount rate. The second possibility is based on the expected value of the uncertain future cash flows. The expected value is then discounted with a risk-adjusted discount rate. Both concepts have certain difficulties. To calculate the certainty equivalent, it is necessary to know the decision-maker's utility function. Instead the second approach requires a calculation of the risk-adjusted discount rate that in turn depends on the chosen strategy.

Thus, an important ingredient for obtaining the optimal decision is a detailed modelling of the decision situation with all future decisions and uncertainties. The main advantage is that decision tree analysis allows consideration of different types of risks as well as managerial flexibility explicitly. However, it remains important to evaluate probabilities for the events at event nodes. This can be challenging, especially for private risks such as technological risks. Expert interviews and experiences from past development projects can help to assess the chances and risks that a development will successfully lead to a new, innovative product. An alternative to explicitly considering managerial flexibility in uncertain decision processes that utilizes market expectations about certain risk factors is the real option approach, which will be discussed in the next section.

2.3 The Real Option Method as a Tool for Improving the Valuation of Research and Development Projects

The previous sections showed that innovative projects in research and development usually provide diverse managerial flexibility while the project is conducted. Management can choose from several potential actions to react to emerging information. These flexibilities can be viewed as real options for management. While classical valuation tools assume that there is no managerial flexibility after the initial investment decision has been made, research and development projects typically consist of a set of real options that can be exercised during the lifetime of the project. The managerial flexibilities are options that can be exercised, but they need not necessarily be exercised. For example, only if the success of the development seems to be threatened, or new information shows that the product cannot be successful in the market, the company would exercise the option of abandoning the project and development efforts. These characteristics suggest the application of option valuation techniques to value managerial flexibility in innovation processes.

Financial option valuation is a well-developed tool, heavily used in the finance sector. A financial option is a right usually without an associated obligation to buy or sell a specified asset for a predefined price (see Trigeorgis 1996, p. 69). Options with the right to buy a certain asset are called call options whereas the right to sell a certain asset is called a put option. The specified price to buy the asset is called the exercise or strike price. Options are traded on all imaginable underlying assets such as common stocks, stock indexes, commodities, foreign currencies, corporate liabilities and so on. In 1973, Fischer Black and Myron Scholes, and later Robert Merton in a generalized version, published a simple method of valuing derivatives such as financial options (see Black and Scholes 1973; Merton 1973). Merton and Scholes were awarded the Nobel Prize in Economics in 1997 for their findings on option pricing theory. The basic idea behind the theory of option valuation is that the payout structure of an option in any state of nature is exactly replicable by a certain portfolio of assets. Then it is easy to calculate the value of the option by simple arbitrage arguments. The replicating portfolio must have the same value as the option order to avoid risk-free arbitrage profits because it provides the same future returns.

To apply financial option valuation techniques to valuing real options it is necessary that an underlying asset with a market price exists, which has the same risk structure as the option. This means that the market-valued asset must approximate the value of the project in each state of nature sufficiently exactly. For market risks in development projects, such assets exist per definition. For such risks, the application of valuation techniques for financial options can be a powerful instrument for valuing real options. For private risks such as technological risks, on the other hand, such assets do not exist. It is hardly imaginable that there might be an asset with a performance that mirrors the probability of the technological success of a product development process.

There is a wide range of possible managerial actions in research and development projects. Here the most important of the numerous real options that can occur in such projects are highlighted. The most drastic decision is to fully abandon the development project. The respective real option is called the option to abandon. It can become advisable to exercise the option to abandon if the probability of a successful development becomes uneconomically low. Another reason may be new market information or a development in the market that makes it unlikely that the new product will be successful. Though this is the most drastic managerial action, it is one of the most frequent occurring real options, which is also regularly exercised in practice. Prior to each stage of a multiple-stage investment project in research and development, it is possible—and due to the gradually resolving uncertainty—highly advisable to review the project and its continuation. This is a direct consequence of the multiple-stage structure of investment projects in research and development.

If it is not advisable to immediately and fully abandon the project, it can still be advisable to delay the investment project for a while. It may be worth exercising the option to delay if, for example, more information about market development is needed. A similar option is the option to defer. This option refers to the optimal point in time at which to start a research and development project. Not only can the decision if and when a project is contracted be made during the development phase but the scale and the intensity of development effort can be influenced. Management has the option to expand and the option to contract the corporate development effort for a certain project.

Real option analysis has helped to highlight the numerous types of managerial flexibility that appear in research and development projects as well as many other corporate multi-stage investment projects. It enables a smart and easy valuation on the assumption that the market is complete in the sense that every risk can be perfectly reproduced by marketable securities (see Smith and Nau 1995, p. 804). This assumption, however, is true only for market risks by definition. But if risks cannot be fully hedged by marketable securities, one can still use decision tree analysis to include private risks and managerial flexibility in the decision process. The relationship between decision tree analysis and the real option method has been extensively researched in academic literature (see Smith and Nau 1995, p. 804). Smith and Nau (1995) showed that on the assumption of complete and perfect markets, the decision tree analysis and the real option approach lead to equivalent results. Trigeorgis and Reuer (2016) provide an overview on real options theory in strategic management.

Usually a decision-maker will have to face private risks and market risks. Smith and Nau (1995) suggest an approach that combines the advantages of real option valuation and decision tree analysis. They therefore set up a combined decision and state tree. Each event node can represent either a private or a market risk. Hence, the first step is to clarify whether a certain risk is private to the innovating firm or whether it is a market risk. After setting up the decision tree, it will be resolved backwards from the terminal nodes of the tree. For each event node with a private risk the certainty equivalent for the node is calculated based on the utility function of the decision-maker and the assessed probabilities for the different states. The value for a chance node that refers to a market risk is calculated based on real option valuation techniques. Thus the hedging portfolio is calculated and the value of the respective portfolio is discounted by the risk-free discount rate. For each decision node the value-maximizing decision is chosen. It is still unclear how the utility functions can be determined. Smith and Nau suggest applying the assumption of risk-neutrality, at least for large publicly held companies because the owners should be broadly diversified and, hence, indifferent concerning the private risk of a single company (see Friedl 2003; Smith and Nau 1995, p. 808).

2.4 The Stage-Gate Process to Manage Flexibility in Research and Development Projects

Once the initial project and investment decision has been made, project structuring, management and controlling becomes crucial to ensuring the project's success. The main goal is a project management that enables a fast and friction free development process, risk avoidance and minimization, the monitoring of the progress that the project is making, and finally, the monitoring of the identified risk factors for the project. Managerial flexibility during the development process can only be utilized if the project plan allows the management to consider certain decision opportunities, and forces the systematic collection and exploitation of information that emerges during the development phase.

A helpful tool in managing innovation processes is the Stage-Gate Process proposed by Cooper (2002, 2008) and Granig (2007). The Stage-Gate Process is a systematic scheme for managing a research and development process. The important process steps (stages) are marked by pre-defined gates. Each gate is connected to an explicit abandonment or continuation decision. Only if the determined criteria are in favor of a project continuation, will it be continued and further resources allocated to the project. The method aims at shortened innovation processes, optimal resource allocation and risk limitation (see Granig 2007, p. 24).

Figure 2 shows an example of a stage structure for innovation processes (see Granig 2007, pp. 193; Cooper 2002): At the first stage the scope of the project is determined. Technological advantages of the project are analyzed and a rough

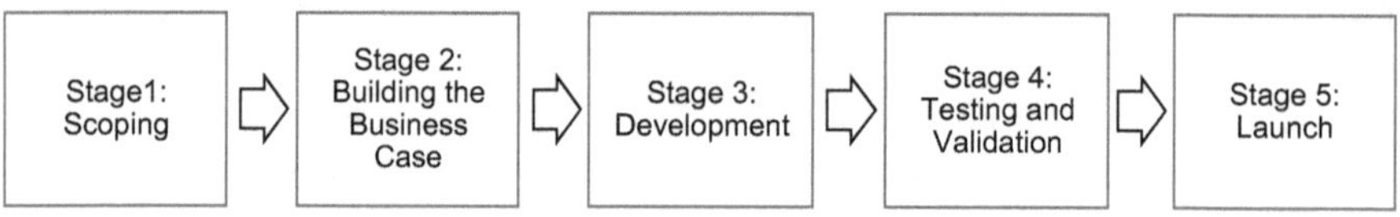

Fig. 2 Stages of an innovative process (Source: Own illustration based on Granig 2007, pp. 192)

market analysis is conducted. Following this, at the second stage, the business case is built. The product specifications are defined and a detailed market analysis is conducted. If the chances of economic success and the evaluation of the financial analyses at the second gate are positive, the third stage begins. Here the actual development of the innovative product takes place. Usually some of the ex-ante uncertainties regarding the technological feasibility and the market success of the product are revealed in this phase. At the following gate, the realization of the planned product and its specification is decided. Further, the economic data about a potential market success are reviewed for a later market launch. At the fourth stage, the product is tested and validated. The product should be brought to market maturity here.

Comprehensive testing is conducted to avoid failure costs later. The final gate prior to the market launch at the fifth stage focuses on the results from the testing and validation stage. If all criteria regarding the product, its specification, and the respective target market and financial analyses are fulfilled, the product is launched at this final stage. The stage-gate method underlines the multiple-stage character of research and development investments. Each gate offers the explicit chance to abandon the project if new information recommends the abandonment. Only the investment expenditures that are necessary to perform the next stage are approved at each gate. Based on our discussion on characteristics and valuation tools for research and development, in the subsequent section we will analyze Siemens' decision to develop its H-class Gas Turbine. We will also describe the management of the development process.

3 Case Study: Development of the H-Class Siemens Gas Turbine

In this chapter we describe and analyze the decision-making process at Siemens for the development of the H-class Gas Turbine. The chapter starts with an overview of energy markets. We especially focus on the role of gas-based electricity generation for the worldwide power supply. The chapter proceeds with an analysis of the market for gas turbines, the global players, clients, and market development. Section 3.2 shows the historic development of Siemens gas turbines. Section 3.3 analyzes the development decision regarding the Siemens H-class Gas Turbine. It starts with an analysis of the strategic reasons for the development of a new generation of gas turbines and the goals that Siemens pursued in the development of the H-class Gas Turbine. We describe the different forms of analysis Siemens did to substantiate the investment decision. We also review the arguments that lead to the decision in favor of the development project. The analysis proceeds by showing the significant risks and uncertainties that had to be faced by Siemens when the decision was made. We describe the project plan and multiple stage structure of the investment that had been chosen to handle these risks and to be able to react to new information during the development phase. Finally, we show how Siemens managed and controlled the actual development process and provide a prognosis on the market success of the H-class Gas Turbine.

3.1 Overview of Energy Markets: A Snapshot, Trends and the Importance of Natural Gas in Power Generation

Between 1991 and 2015 worldwide power generation almost doubled. In this period it increased from 12.106 TWh in 1991 to 23.208 TWh in 2015. In particular, emerging regions such as South America, Asia and the Middle East have seen a fast-rising power demand and there is still a steadily increasing and worldwide demand for reliable, flexible and cost-effective power generation (Source: IHS Energy 2016, Rivalry scenario). Scarcer resources and an increasing awareness of environmental pollution and climate change demand more efficient, flexible and sustainable power generation. The typical energy sources that are used for power generation are coal, natural gas, nuclear power, lignite, wind, water, and other regenerative energy sources. Figure 3 shows the worldwide distribution of energy sources for power generation in 1990 and 2015.

Since 1990, the share of natural gas in worldwide power generation has increased from 14.8% to 23.0%. In absolute numbers, this is an increase from 1756 TWh per year to 5488 TWh per year (Source: IHS Energy 2016, Rivalry

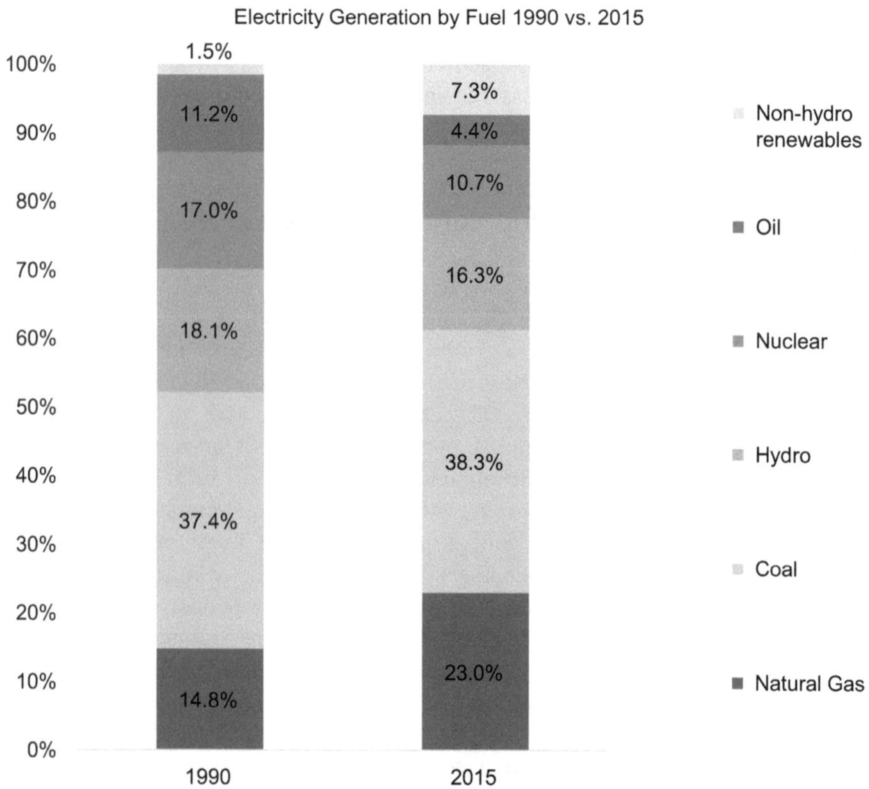

Fig. 3 Electricity generation by fuel, 1990 and 2015 (Source: IHS Energy 2016)

scenario). In Europe, this trend is even more significant: Since 1990 the share of natural gas in European power generation rose from 7% to about 17% in 2015 (Source: IHS Energy 2016, Rivalry scenario). The share of gas in electricity generation differs widely between countries even within Europe. The Netherlands, with the highest share in Europe, generates more than 50% from natural gas whereas Sweden generates almost no power from natural gas. Germany produces about 10% of its electric power from natural gases. Until 2030, the annual total of worldwide electricity generation is expected to rise to 34,400 TWh. This implies an annual growth rate of 2.5%. Further, the share of gas in total electricity generation is expected to increase to 24% of the total worldwide electricity generation. The largest rise is expected to be seen in the share of renewables electricity generation, from about 4% in 2011 to 17% in 2030.

The importance of combined cycle power plants is currently significantly increasing, but the reasons for this trend are manifold and vary for different global regions. In the United States in particular the price for natural gas has fallen sharply since hydraulic fracturing has offered a new, comparably cheap way to drill for natural gas. This price decrease makes it economically more attractive to obtain electricity from natural gas. In the medium run, new means of gas production will also lead to a certain extent to a decoupling of the gas price from the oil price that can make gas as an energy source for electricity generation even more viable. In countries with very low gas prices, combined cycle power plants can even contribute to base load capacity.

By contrast, for example, in Germany, base load coverage with natural gas is uneconomical due to the high gas prices in Central Europe. Whereas the average price in 2016 in the United States was USD2.5 per MMBtu, in Germany it was USD4.4. Thus, gas-based electricity generation is typically used for medium and peak load coverage in Germany. Short start-up times are vital for a power plant to be suited for coverage of peak loads. It is usually possible to rev up a gas-fired, open cycle plant within a few minutes. With a rising proportion of renewables generated power the importance of gas power stations might even increase. With a traditional, mainly fossil fuel electricity production mix, the main reason for unexpected need for additional capacity was occasional peaks in the power demand.

With a rising share of regenerative sources, a second source of unexpected need for additional capacity appears: It is very hard to predict whether the sun will shine or whether and with what speed the wind will blow at a certain time of a certain day. Thus, in a network relying to a high degree on energy from sunlight and wind it will be necessary to get quickly available power when energy from regenerative sources experiences unforeseen changes within a short time. Though combined cycle power plants can solve this problem, there is an inherent economic problem. If subsidized regenerative energies are available during a relatively large number of hours during the year, gas turbines run only during a relatively low number of hours, even though they have an important function in guaranteeing a reliable power supply. However, it can become inefficient for power suppliers to invest in combined cycle power plants if they rarely operate during the year. These trends also increase the pressure on OEMs of gas turbines to reduce the investment and operating expenses for power suppliers using gas turbines.

In the long run, extended transport capacities might lead to more trade of gas between different regions of the world and thereby less significant differences in prices. Thus, gas prices in different regions of the world might converge, leading to lower prices in Europe and the Far East. Trade is currently restricted by fully utilized transport capacities. To sum up, the worldwide power markets are expected to further increase, and due to several global trends make it likely that the share of gas in total power supply will also rise. In the subsequent section we take a closer look at the market for gas turbines.

3.2 Gas Turbines: History and Market Analysis

The H-class Siemens Gas Turbine is the latest generation of gas turbines developed by Siemens. Gas turbines are used in power plants to generate electricity from natural gas. Today the large, heavy duty gas turbines are usually installed in combined cycle power plants. In a combined cycle power plant, natural gas is used to drive a turbine generating electricity by a generator. The hot exhaust gas from the gas turbine is utilized to evaporate water for a downstream steam turbine. Using this combination, the highest efficiencies can be reached because unlike a pure gas power plant, the energy from the hot exhaust gas is not wasted. The launch of the H-class Siemens Gas Turbine development project dates back to October 2000. After a 5 year development phase, followed by 2 years manufacturing, and 1 year installation, the prototype of the gas turbine was first employed in a power plant in Irsching for validation purposes from 2008. In 2011, the extended combined cycle power plant in Irsching went into operation. The first commercially ordered turbine was inaugurated in 2013 in Florida. It is the first gas turbine with a combined cycle net efficiency of over 60%, which is an increase in efficiency of about 1.7% points compared to the predecessor, the Siemens F-class Gas Turbine. Besides the outstanding efficiency rate, a high degree of flexibility via short start up phases and high part load performance, high reliability, and reduced life cycle costs are the main characteristics of the new generation of gas turbines.

The Siemens H-Class Gas Turbine falls under the category of large gas turbines (LGT). LGT are gas turbines with power of more than 60 MW. The average unit price for large gas turbines is about EUR 20 million to EUR 50 million depending on the output sizes and the contract volume. That comprises the cost of the turbine, the generator, the process control technology and the side systems. There are currently four original equipment manufacturers (OEM) for large stationary gas turbines with a capacity of more than 60 MW: General Electric (USA), Siemens Energy (Germany), Mitsubishi Hitachi Power Systems (Japan) and Ansaldo (Italy). Alstom Power (France), another manufacturer of large gas turbines, was acquired by General Electric in 2015.

In 2012, Siemens had an overall market share by ordered capacity of 28% of the whole market for gas turbines. The market leader was General Electric with a market share of 43%. Mitsubishi Hitachi Power Systems was third with a market share of 18% and Alstom followed with 4%. The global market ordered a total capacity of 53,429 MW or 585 gas turbines including smaller and mid-range gas turbines. The ranking of the four market players has been stable over the last few years. Between 2007 and 2012 the market share of General Electric ranged between 40 and 48%, Siemens between 24 and 39%, Mitsubishi Hitachi Power Systems between 7 and 18% and Alstom between 2 and 11%. However, in some years Siemens could catch up with its main competitor General Electric. For example, in 2009, General Electric had a market share of 41%, compared to Siemens with a market share of 39%. Figure 4 shows the development of market shares between 2007 and 2012.

Following with the financial crisis, the gas turbine market has seen significant fluctuations in demand over recent years. Shortly before the crisis began in 2007 a total capacity of 82,294 MW was ordered. This rapidly fell to a low demand in 2009 when only 46,011 MW were ordered. Though the market recovered slightly, one could still observe large fluctuations in the following years with an ordered capacity of 74,675 MW in 2011, and, as stated above, again only 53,429 MW in 2012. Low gas prices and a higher demand for flexible power production offer good prospects of a positive market development for gas turbines. The research firm Forecast International estimates that the market volume for the next 10 years will be 12,000 gas turbines worth about EUR 168 billion. A market study by IHS Energy from 2016 forecasts a solid market growth for the global gas turbine market until 2030: globally new installations of gas fired power plants grow from 46 GW in 2015 to 71 GW in 2030 (Source: IHS Energy 2016, Rivalry scenario). Clients for large gas turbines are worldwide power plant operators. Countries with the largest

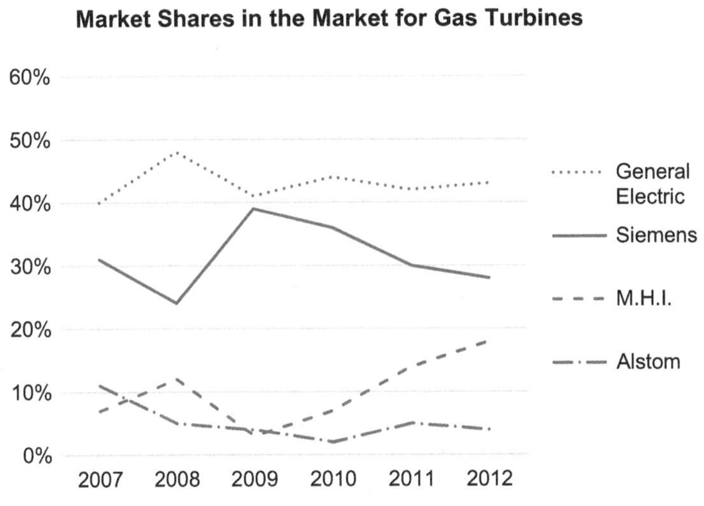

Fig. 4 Development of market shares measured in MW of delivered capacity (Source: McCoy Data)

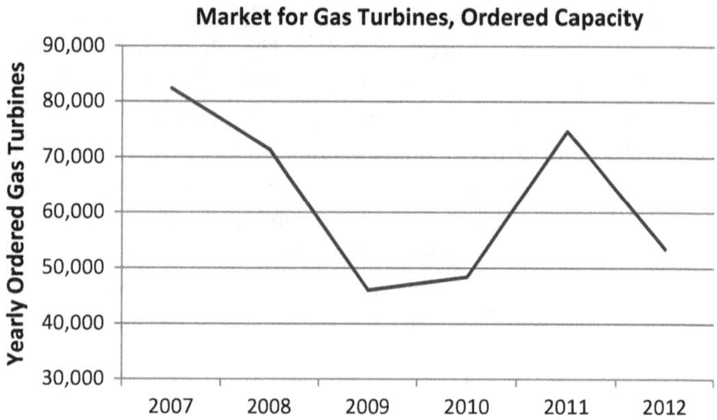

Fig. 5 Development of the market for gas turbine (Source: McCoy Data)

orders of gas turbines have been the United States, Saudi Arabia, Russia, China and Japan. With the exception of Japan, these are countries with domestic gas production and as a consequence at least moderate gas prices. However, about the half of the orders go to other countries (Fig. 5).

There are two typical project constellations for gas turbines. In the first constellation, the OEM only delivers the turbines for the power station. In this case, a third-party supplier does the construction and installation of the power station. In the alternative constellation, the OEM will also take over the construction and installation as general contractor. In addition to the sale of gas turbines, and possibly the construction and installation of the power station, service and maintenance is an important source of income for the OEMs. For the first few years, the OEM usually concludes a full-service contract with the client. Depending on the project's specification, these service contracts can have a volume of up to 50–70% of the initial price of the gas turbine or power plant.

As we have already described above, gas and steam turbine power stations are an important component of the worldwide power supply. The first operation of a gas turbine in a power station dates back to 1939, when a four MW gas turbine from BBC Brown were utilized in an emergency power station in Neuchatel in Switzerland. In the following years, the most important application of gas turbines was in jet engines, where it remains an important engine type until today. However, with the development of combined cycle power plants and increasing efficiency, gas turbines were increasingly used for electricity generation after the 1970s.

In recent decades, gas turbine series have been newly developed usually every 10–20 years. The Siemens E-class Gas Turbine, with a capacity of 150 MW, was developed in the seventies the F-class Gas Turbine, with a capacity of 250 MW, in the nineties. The series are mainly distinguished through their power range and the efficiency of the turbines. In a combined cycle power plant, the E-class Gas Turbine reached 450 MW with two gas turbines and an efficiency of about 50%. In its first version the F-class Gas Turbine reached a combined cycle efficiency of 56% with

705 MW, again with two gas turbines of a combined cycle power plant. For power plant operators as main clients for gas turbines, the power range and the efficiency are the most important characteristics of a gas turbine especially with respect to the operational expenses during the lifetime of the power plant.

Having reached the 55% mark of combined cycle efficiency, it appeared to be clear that the next technological step was the 60% mark of efficiency. All internationally renowned manufacturers strived for the 60% mark of efficiency. In the United States, the Department of Energy (DOE) funded the Advanced Turbine Systems Program (ATS Program), in which besides the market leader General Electrics, the second largest manufacturer of gas turbines in the US, Westinghouse, also participated with its unit Westinghouse Power Generation. From the middle of the nineties the competition in the power plant market continued to intensify. Worldwide, dramatic changes in the market could be expected driven by further liberalization and privatization of energy markets. Within this environment, Siemens wanted to strengthen its market position. In 1997, Siemens decided to buy its competitor Westinghouse Power Generation for converted EUR 1.33 billion. Westinghouse Power Generation employed 8000 people and had converted sales of EUR 1.9 billion in 1996.

With the takeover of Westinghouse Power Generation, the Siemens business unit for energy generation, the former Siemens KWU, became the second strongest player in the global market. The takeover opened up additional markets for Siemens power plants and gas turbines. Westinghouse Power Generation in particular had a strong position in the United States and Saudi Arabia. In summary, Siemens acquired well-developed technology, attractive market access, and grew significantly. On the other hand, after the takeover different technological concepts and solutions existed simultaneously in the same company. Most of the components for similar turbines were now available in two varieties. This increased the complexity in the Siemens gas turbines section significantly and caused additional costs.

This was one among several reasons why Siemens considered developing a completely new generation of gas turbines 10 years after the previous generation had been introduced. Technical evolution seemed to enable more modern gas turbines that better met the market demand and offered added value for customers. It had been 10 years since the introduction of the latest generation and, thus, the length of historic product life cycles suggested considering the development of a new generation of gas turbines. Additionally, General Electric had already announced to develop an H-class gas turbine. This announcement further raised the question whether Siemens also need to develop a larger and more efficient generation of gas turbines to remain competitive. A second argument specific to Siemens at the end of the nineties stimulated the impetus to create a new generation of gas turbines. The development of a new gas turbine generation would offer the opportunity to integrate Siemens and Westinghouse concepts into a common family and thereby lower the complexity resulting from the takeover of Westinghouse Power Generation.

Due to the long product life cycles and exceedingly high development costs, a generation of gas turbines is usually technologically upgraded several times during its lifetime. Individual components are developed further to maintain a technologically competitive product. Accordingly, the predecessor, the Siemens F-class Gas Turbine went through several upgrades: The initial version was, for example, used in 1996 in a combined cycle power plant in Didcot (United Kingdom). It reached an efficiency of about 56%. A later upgrade was used in a combined cycle power plant in Mainz in 2007 and reached an efficiency of over 58.5%. This improvement was reached without a general reconstruction of the gas turbine. In contrast to a technological upgrade of an existing generation of turbines, the development of a new generation means development from the ground up. In a later section, the most important technological decisions for the new gas turbine will be discussed.

To sum up, we identified several reasons for the development of a new gas turbine generation at Siemens. However, a resource-intensive and risky project such as the development of a new gas turbine generation must be carefully analyzed. The next section describes important elements of the decision-making process for the development of the H-class gas turbine at Siemens.

3.3 The Decision Process for the H-Class Gas Turbine

The development of a new generation of gas turbines is a highly resource-intensive project. Furthermore, the development process is extremely complex and is accompanied by many uncertainties. Thus, a careful decision and valuation process was necessary to enable Siemens to make an informed decision. The risk of losing out to the technological development of its competitors and the market-side chances of a new generation had to be traded-off against the investment expenditures that were necessary for the development process, and also the manifold risks that would affect the development decision. The investment decision and the whole development process were structured by a strict project plan divided into five distinct phases. The first phase was strategic product planning, which included the product strategy as the first step. In this planning phase, the main goals for development should be set and the initial development decision will be made. It comprises a definition and technical specification of the product to be developed as well as comprehensive market and risk analyses. The following paragraphs show the important steps and analyses that were conducted to reach the decision. Afterwards, the complete development process plan is outlined in greater detail, and the development process will be discussed.

Main Goals of the H-Class Gas Turbine Development

The first step in the product strategy phase of the decision process was to define the main development goals as precisely as possible. Firstly, this refers to technological specifications of the turbine generation that should be designed. It was also crucial to fix goals for the economic savings for potential buyers of the new H-class gas turbine. The liberalization of the energy markets, the privatization of the energy sector in many countries, and the increasing market penetration of renewable energy resulted in more complex customer requirements (see Fischer and Nag 2011; Ratliff et al. 2007). Firstly, liberalized energy markets increased the price pressure on the electricity generating industry. Therefore, a more cost-efficient power production was key for the power generating companies and increased their demand for efficient turbines and power plants. To meet this demand, Siemens aimed to reach a combined cycle net efficiency of over 60% that should result in about 3% of fuel savings. Furthermore, the specific investment costs (investment expenditure per produced kW power) had to be lower than for the predecessor.

Besides the savings in investment and operation expenses, the development of the H-class Gas Turbine aimed at a higher flexibility for the turbine's operation. The gas turbine alone should be at full power within <15 min. This is crucial if a combined cycle power plant is to be able to cover peak-loads that more frequently appear in electricity grids with a high share of renewably generated power. Moreover, for a highly flexible deployment of power plants, outstanding part load behavior is vital. In many countries companies are forced to lower their emissions. To address these requirements the newly developed turbine should have significantly reduced emissions per kWh produced. As outlined in Sect. 3.2, the service and maintenance of gas turbines plays an important role. A basic inspection is necessary every year, supplemented by several major revisions every 3–6 years. The relatively high frequency of inspection makes it necessary to allow for a serviceability with short outage times, which was defined as a further goal of the H-class turbine's development.

All in all, the new generation should minimize life cycle costs to increase the net present value for the power plant's owner. From the very beginning Siemens defined clear and highly ambitious aims. To achieve market success, the life cycle costs should be lowered by at least 7–8% compared to the previous gas turbine generation.

Market Analysis for the H-Class Gas Turbine

The next step for Siemens was to analyze the potential market for an H-class gas turbine, and to forecast how this market might develop over the next one or two decades. As in the whole period from 1991 to 2015, in the years prior to the decision by Siemens, the worldwide power supply grew by about 3% annually. Even more important are the growth rates for the years after the projected market launch of the new gas turbine generation. For the years 2015–2030 an only slightly slower growth in worldwide electricity generation was expected. For this period forecasts predicted an average annually growth of 2.5%, from 23,208 TWh in 2015 to 34,400 TWh in 2030.

It was not only the absolute amount of worldwide power supply that was an important factor for Siemens. Perhaps even more important for the decision was the forecasted development of gas-based electricity generation. This was expected to grow even faster than the total electricity generation. For potential sales number after the market launch, a key factor is how gas-based electricity generation will grow after the development is completed. Figure 6 shows a forecast for the development of shares in world electricity generation. Though renewables are

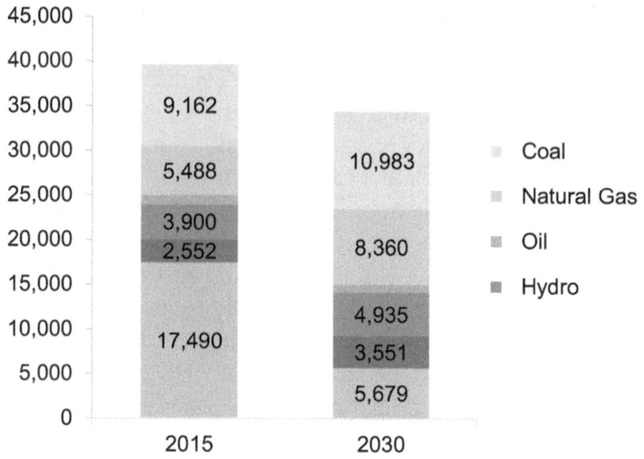

Fig. 6 World electricity generation (TWh) between 2015 and 2030 (Source: IHS Energy 2016, Rivalry scenario)

gaining in importance, there is still a forecasted increase in the share of gas-based electricity generation of 23–25% between 2012 and 2030. All in all, the development of the worldwide power supply suggested that there would be a promising market for a new class of gas turbines for Siemens.

Subsequently, Siemens went deeper into the detail and forecasted which capacity would be installed in 2012, which capacity should be installed in 2030, and which retirements of power plants are expected in the intervening period. With these data it was possible to calculate which new capacity additions would be necessary in the two decades after the projected market launch of the new gas turbine generation. Here we focus on the development of combined cycle power plants as the most efficient way of using the large gas turbine. The installed capacity in 2015 was 1605 GW worldwide. The expected development of the energy mix as it was described above leads to a forecasted worldwide installed capacity in combined cycle power plants of 2204 GW in 2030. With an expected retirement of 222 GW of installed capacity, one obtains new capacity additions of 821 GW between 2015 and 2030.

The revenues from a gas turbine or the whole combined cycle power plant greatly depend on the project specifications. A rough range for a combined cycle power plant is EUR 600–EUR 800 per kilowatt of installed capacity. In a pure gas power plant the price drops to EUR 200–EUR 400/kW installed capacity. Siemens' market share in the market for gas turbines was about 25% at the beginning of the past decade. Assuming a constant market share for Siemens, this would result in nearly 205 GW combined cycle power plants for Siemens out of a total capacity addition of 821 GW in the years between 2015 and 2030. These potential orders would equal a volume of EUR 144 billion. The development of a technically superior new generation of gas turbines might enable an even higher market share, and thereby even raise the potential market volume for Siemens.

Even though there clearly would be a market for modern gas turbines, and Siemens would be the first, after the failed attempt by General Electric in Baglan Bay with a gas turbine as large and as efficient as the planned Siemens H-class Gas Turbine, it was not certain whether market participants really would pay for such a large gas turbine. A market analysis reviewing the past and trying to forecast the future sales figures for different types and sizes of gas turbines in the market was conducted. The development as well as the forecast confirmed that there is a tendency towards larger gas turbines. All in all, Siemens concluded that the market for gas-based electricity generation is growing probably even faster than worldwide energy generation, which is expanding at a rate of around 2.5%. A significant number of new power plants using gas turbines would be required in the years following the projected market launch, and there seemed to be a market for larger and more efficient gas turbines. The market volume for gas turbines only in combined cycle power plants promised to be triple-digit billions over the lifetime of the new gas turbine generation.

However, in order to analyze the profitability of the development project, the forecasted revenues are traded-off against the expected investment expenditures for the project. An ex-ante estimation of the total investment expenditures is very complicated. A rough estimate of the investment expenditures for the development of the gas turbine's 50-Hz-version was EUR 150 million. Another EUR 50 million were necessary to develop the 60-Hz-version. These numbers did not include any expenses for the very resource-intensive testing that would definitely be necessary after the development were completed to avoid nonconformance cost, which could be very high as they are usually contracted as a percentage of the overall contract for the power plant. The expected market success of the H-class Gas Turbine and the potential revenues would, however, justify these investment expenditures, but a careful analysis and assessment of all potential risks and uncertainties was considered crucial for the success of the Siemens project. In the following section we will outline and classify the most important sources and types of risk Siemens associated with the development project.

3.4 Identification and Assessment of Relevant Risks in the Project

A wide range of risks and uncertainties had to be taken into account for the project decision and—in the case of a positive project decision—managed. To ensure a systematic and comprehensive analysis of all significant risks in the project, they are classified according to the source of the risk. Here we differentiate market-related risks, technological risks, and other risks, such as political risks. Market-related risks refer to all uncertainties regarding the market success of the new gas turbine generation. Technological risks refer to uncertainties regarding the technical realization and the development process of the new gas turbine generation. The market estimations from the previous section are subject to considerable uncertainties. The most important and uncertain variables are the whole market volume, the market price level, and Siemens' market share. They all depend on a wide variety of influencing factors and variables that are unknown, and somewhat hard to predict. Siemens estimated the market volume by the development of worldwide energy demand and the share of gas-based electricity generation in respect of the total energy supply. Historically, the growth rate of worldwide energy demand shows some minor fluctuations, and indeed the growth rates for the second and third decade of the twenty-first century are forecasted to be slightly lower than they were in the last century.

However, the worldwide energy demand has steadily grown, and with fast-growing economies and the rising wealth in emerging countries, one can expect continuing growth in power demand at least for the lifecycle of a further gas turbine generation. More important, and harder to predict, is the share of gas-based electricity generation. An important factor for the long-term development of gas-based electricity generation is the price of natural gas itself. With rising gas-prices, the gas-based electricity generation would definitely lose in significance, whereas lower gas prices would raise the probability of an increase in the capacity of gas-based electricity generation. However, the prediction of actual gas price development is in general very hard. Besides gas price development, there are other important factors determining the importance of gas-based electricity generation. Of course, the price for other energy sources is another important factor. Increasing prices for coal or oil would make gas-based electricity generation economically more attractive. The development of more efficient power plants with other fuels, however, would make gas as an energy source less attractive. Regulatory changes with the goal of reducing fossil fuels that may also affect gas-based electricity generation might also be more likely. For example, in Germany regulatory changes influenced the economic profitability of gas-based power plants.

As described in Sect. 3.1, the unlimited priority feed-in and the subsidies for regenerative energy sources make the operation of gas-based power plants economically unprofitable during a large number of hours per year. The respective power plants are hard to finance with the remaining number of hours, where the price for electricity is high enough for economic operation of gas-based power plants. In summary, there are a lot of very different factors that influence the development of the gas-based electricity generation, such as the price of gas and its potential substitutes, the development of alternative technologies and the regulatory environment. Finally, with respect to the market volume and planned sales figures, one has to be aware that the market for gas turbines is usually cyclical in nature, with business cycles of between 2 and 5 years.

The development of the share of gas-based electricity generation on total power supply mainly determines the need for new gas-based power plants. However, besides the market volume for gas turbines, there is still uncertainty about the market share that Siemens can achieve. With the successful development of a new larger gas turbine and a combined cycle efficiency of more than 60%, Siemens would clearly achieve a first mover advantage that might ensure a stable or growing market share. If Siemens achieves the targeted cost savings for clients of 7–8% of lifecycle costs, it offers a competitive and attractive product. However, as described in Sect. 2.4, every important competitor was known to be aiming at the development of a more efficient large gas turbine breaking the all-important 60% combined cycle efficiency barrier. But such developments would make it even more important to develop an attractive product to stay in competition, and to mitigate the risk of being left behind the competition tomorrow.

The gas price is determined as an important risk factor for the market development of gas turbines. Whereas a low gas price is favorable for a high total market volume, this effect might be somewhat compensated by a countervailing effect on the market share for the H-class gas turbines: A high gas price increases the value of highly efficient gas turbines. The last important factor for the potential revenues from the new generation of gas turbines is the market price levels of gas turbines. In the past, gas turbines had been seen to be subject to strong price fluctuations of up to 30% in both directions around the long-term average.

The major uncertainties with respect to the market success of the new generation of gas turbines stems from the development of gas prices, its substitutes, changes in the regulatory environment, and the actions of the main competitors. The critical factor in ensuring a successful market launch generally is the technological success of the development process. Thus, besides the analysis of market-related risks, Siemens carefully analyzed the technological risks that would be related to the development of the new Siemens H-class Gas Turbine. Firstly, one can differentiate between technological risks during the development phase and risks after the market launch. First of all, it was uncertain whether the planned duration of development and thereby the planned investment expenditures could be realized. Large development projects tend to last longer than expected or become more expensive than planned. Furthermore, it was not sure whether the technical and economical specifications of the new gas turbine generation could be met. Siemens set itself strict and ambitious development goals. If for example the efficiency goals could not be met or their realization became more important in terms of development expenditures or production costs, the gas turbine can easily become less attractive to the market and, thus, less successful than expected. This shows that there are close interdependencies between the technological and the market-related risks.

There is a second category of technological risks that appears after the market launch of the product. If a new technology is not sufficiently tested and proven, there is a significant risk of high losses after the market launch. Product failures, necessary repair works and machine downtimes can result in high costs for the manufacturer and the client. Siemens and other OEMs had such experiences after the launch of the F-class gas turbines in the mid-nineties. The total failure costs amounted to many billions for manufacturers and power plant operators, and thereby exceed the original development costs many times over.

The only possibility of avoiding such excessive failure costs is an intensive testing phase. This is of particular importance because some components in gas turbines are too large to test separately before a prototype of the whole gas turbine is developed. The single construction and manufacturing of a large gas turbine is a very expensive project especially together with a whole power plant that is necessary to fully test a prototype gas turbine. The power plant together with the gas turbine can cost far more than a hundred million EUR. Among other reasons, due to these high costs OEMs usually sold a prototype to a power plant operator where it was used and simultaneously tested. The operator could sell the turbine's power with a discount, and in return the OEM obtained permission to run further tests with the sold turbine while it was in operation in a power plant. However, this frequently caused different problems. The major problems were conflicts about when the turbine could be tested while it was in operation. The power plant operator relied on a steady operation of the turbine whereas the OEM needed the turbine for tests and verification at certain points of time. Siemens deviated from this approach and tried to find a better solution to the testing phase of the H-class Siemens Gas Turbine. The stated aim was a comprehensive testing phase to approve the new technology and thereby avoid later failure costs.

Siemens decided to build a prototype after the development phase that should not be sold to a power plant operator right away. Instead Siemens entered into a cooperation agreement with the German plant operator E.ON. Together with E.ON, Siemens planned to build a gas-based power plant in Irsching, where the prototype of the new H-class Siemens Gas Turbine was to be tested over 18 months. During the testing phase the power plant is to be under the sole authority of Siemens. Of course this would cause further testing costs but it allows Siemens to extensively test the new gas turbine generation without having to take the ongoing operations of the power plant for the operator into consideration. After the 18 month testing phase, the power plant is to be expanded to a combined cycle power plant and afterwards it should be sold to E.ON. Hence, Siemens should be able to earn back a part of the investment expenditure for the testing power plant.

Risk Assessment for the Development Decision

Siemens used a wide range of analyses to evaluate the profitability of the development project, and the effect of the risk factors described in the previous section. This section gives an overview of the most important analyses. The key target figures that were considered during the analyses are the forecasted quantities of gas turbines to be sold, the profit margin, development expenditures, break-even numbers, and the time to break-even. Siemens conducted scenario analyses to analyze the effect of the gas price on the market volume, and thereby the product's success. Further, different scenarios for the competitors' behavior and the development of the competition were applied. Additionally, sensitivity analyses were applied. With sensitivity analyses, the effect of the market share, the market price level, and changes in the gas price on the profitability of the project were calculated. Siemens calculated threshold values of these primary variables for the time required to break-even.

The gas price has been identified as one of the most important market influences on the success of the planned gas turbine. The previous section has shown that high gas prices have a negative effect on the market volume, but can also have a positive effect on the market share of a highly efficient gas turbine, as the H-class turbine was planned to be. Until now, the gas prices in different regions of the world differ significantly. Due to the increased production volumes by hydraulic fracturing, the price for natural gas in Northern America may be significantly lower than prices in Central Europe. To take these differences in the market situations into account, the market volumes were separately calculated and forecasted for the different regions of the world. The importance of the gas price as a main factor in market development but also the role of increased efficiency in the gas turbine is underlined by the total cost of ownership (TCO) analysis of a gas turbine. On the basis of a 25-years product life cycle, the investment expenditures comprise only about 15% of the TCO. By far the largest part of the total ownership costs stems from the costs for the utilized gas. They can rise up to 70% of the TCO in times of high fuel prices. The remaining 15% are service costs and expenses that occur for the power plant operator. Following these calculations, the effect of an increase in efficiency by 1% can result in reduced life cycle costs of a base-load plant by EUR 30 million to EUR 50 million in markets or times where gas prices are high.

Besides the economic risks and uncertainties, there are substantial technological risks. These range from the risk of not achieving ambitious product characteristics to default costs for clients after the market launch. The risk management process for technological risks in the development process was guided by the Siemens Risk Assessment tool. The first phase of the risk assessment process aimed at identifying all the technological risks. The process started with a workshop that was moderated by process experts. Qualified engineers and experts tried to identify all relevant risks, such as default risks in certain components. It is not possible to completely avoid such risks and defaults. Hence, the risk management process aimed to mitigate the risk. The risks were classified as low, medium and high risks, depending on the risk level. The risk level was determined by the estimated probability of occurrence and the possible loss that can be caused by a default. From the beginning, all analyses aimed not just at the identification of risks but also at quantifying the costs for a default. For example, experts tried to calculate the days that a gas turbine fails in case of the failure of a certain component. Thereby, the monetary losses of such a failure could be calculated. Following particular development steps, the respective risks were regularly updated with new information that was gained during the development. Furthermore, the determined non-conformance costs were integrated into the business case to obtain a business plan that incorporates the technological risks as well as possible.

3.5 The Project Management and Managerial Flexibility in the Development Phase

After considering all relevant aspects for the profitability of the project and potentially related risks, Siemens decided to launch the program with the concept phase on October 1st, 2000. Along with a strict project plan, coherent program management is necessary for the success of a highly complex development program, such as the development of a new gas turbine. The development process plan was comprised of nine steps in five distinct phases, which described the primary functions within the business that had to be met in order to achieve successful developments. The main phases of the development process are strategic product planning, design, sales preparation, design implementation and validation. These phases reflect the main steps of Cooper's Stage-Gate Process (see Cooper 2002). The whole development process was strictly organized as a stage-gate process. Figure 7 shows the structure of the development process for the Siemens H-class Gas Turbine.

Either during or after each of the five stages described above, certain evaluations, called a gate, should be made. At these gates, decision-makers can make important decisions about the development project. Most importantly, the profitability of the project's continuation is reviewed. The first gate follows the determination of the product strategy during the strategic product planning stage. The second stage, the design phase, contains two steps, the conceptual design and the basic design phases.

Following the conceptual design, a second gate was set. During the third stage, sales preparation, the commercialization of the gas turbine was planned. This gate is followed by the third, the product release. The fourth gate should be the design

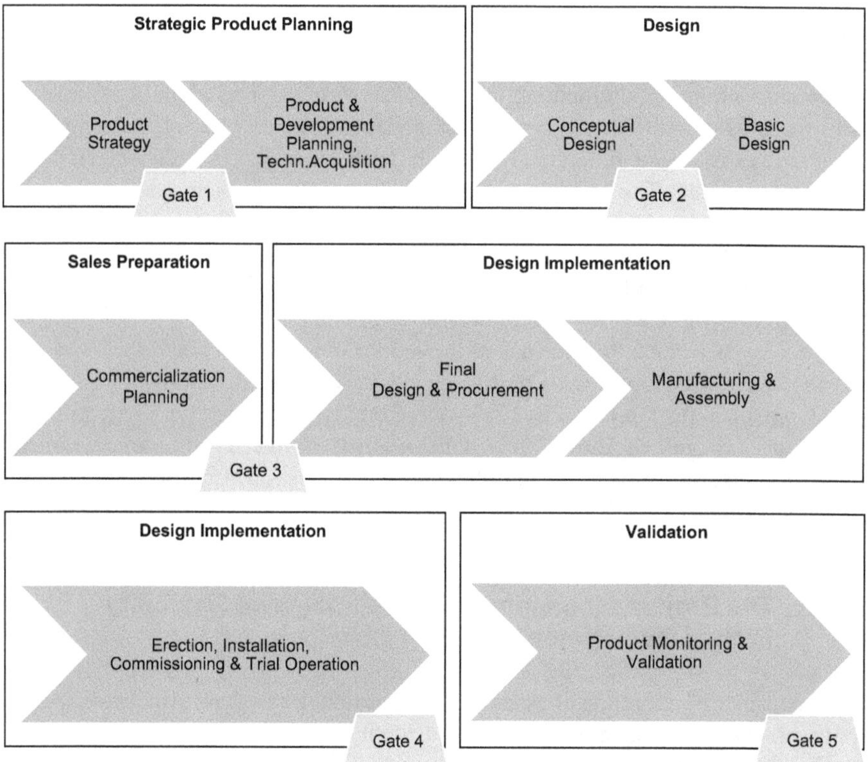

Fig. 7 The development process for the Siemens H-class Gas Turbine (Source: Siemens company data)

implementation. The first step in design implementation was the final design and procurement, followed by the manufacturing and assembly of the product. Design implementation should be completed by the erection, installation, commissioning and trial operation of the new turbine. Afterwards, in the fourth gate the series should be launched. The final stage is validation in which the product's performance is monitored and validated. After the final stage, the last gate signals the completion of the development phase.

At each gate a decision about the continuation of the development project must be made. In so doing Siemens enabled the flexible use of new information during the development process. These critical high-level decisions at predefined stages of the development process should ensure a regular re-examination of the success of the project and the market development. Only after this review and an explicit decision in favor of the project's continuation would new funds for the next development step be released. All of these gates represent options to discontinue the project. The gates are purposely chosen at specific points in the development process where particular information would become available that is important for the decision on continuation: For example, the determination of the product strategy that is followed by the first abandonment decision at gate one provides a preliminary indication of the potential market's attractiveness. For the second continuation decision following the conceptual design gate, there should be more clarity about the technological feasibility, the necessary development effort and the potential market development for gas turbines. After the commercialization planning, more detailed market data for the third abandonment decision should be available. Only if the financial outlook for the gas turbine in development is sufficiently positive would Siemens begin with the actual design implementation. The successful of the design implementation will be monitored for the fourth continuation decision. All mentioned gates entail a real option for Siemens. A growing body of literature analysis real option approaches in the energy and electricity sector (e.g., Fernandes et al. 2011 or Ceseña et al. 2013).

The Development Phase and Technological Flexibility

The H-class Gas Turbine development program was launched with the start of the first stage, the concept phase, on October 1st, 2000. The determination of the product strategy took 6 months and was completed with the product strategy release on March 21st, 2001. During the conceptual design phase which followed, the most important technological decisions were made: For each component all potential solutions were considered.

One of the most important technical decisions to be made was which engine cooling system should be used for the gas turbine. Engine cooling is important to ensure that the components along the path of the hot gas can withstand the high temperatures. Siemens had to evaluate whether steam-cooled technology was better than a completely air-cooled turbine. With a steam-cooled turbine, higher efficiency rates can be achieved. On the other hand, an air-cooled turbine has higher operating flexibility. The steam must be generated before the turbine can commence operation. The necessary generation of the steam leads to longer startup times. Furthermore, a steam-cooling system leads to higher technological complexity and increases the risk of failures. Siemens' main competitor, General Electric, had already begun developing a new generation of gas turbines with the declared target of a 60% efficiency rate. For this development General Electric relied on the steam-cooling technology.

For the previous generation, the Siemens F-class Gas Turbine, a purely air-cooled engine concept was used, while the latest gas turbine that Westinghouse had developed employed a combined air- and steam-cooled approach. In the industry, it was generally considered not possible to achieve such an efficiency rating with a purely air-cooled system. However, there were several future trends that some thought might make a more flexible operation of gas turbines favorable: More independent power producer projects and the expected increase in renewable power generation makes it necessary to have back-up power plants with a high operational flexibility to balance out fluctuations in decentralized or renewable power production. Flexible gas-based power plants are an important component for counter-balancing unexpected changes in renewably generated power. Trends such as decentralized and renewable power generation highlighted the necessity of flexible gas turbines. Additionally, Siemens asked its customers for feedback regarding the importance of high operational flexibility. These arguments ultimately led to the decision in favor of a completely air-based cooling system. Among other-decisions this was fixed at the second gate, which was concluded on November 5th, 2001. With this gate the basic design phase started which was finished with the third gate, the product release, on August 17th, 2004. The decision for the air-cooled engine concept demonstrates how Siemens used technical flexibility to respond to new trends that seemed to be developing.

The Testing Phase

Already in the early planning phase it had become evident that a gas turbine with more than 220 MW, which the H-Class Gas Turbine was designed to far exceed, could not be tested at Siemens' Berlin Test facility (see Fischer 2011). Thus, Siemens decided to cooperate with the large German electricity provider E.ON to launch a combined cycle project, to make it possible to test the gas turbine at the site in Irsching, a real plant environment with a grid connection. The infrastructure for a large combined cycle power plant already existed there. Moreover, it offered the possibility to increase capacity in Southern Germany to compensate for the newly installed wind power capacity in Northern Germany. Thus far, three gas-fired units have been installed in Irsching. Irsching I and II, built in the 1960s and 1970s, are no longer in operation, whereas Irsching III can still be used for peak load operation. As described above, the OEM of the gas turbines usually sold the prototype to a power plant operator and tested it in operation. However, this approach can lead to high breakdown costs and conflicts between the gas turbine manufacturer and the power plant operator regarding testing and operating times. To avoid such problems, Siemens decided to fully test the gas turbine series prior to the commercial market launch.

Siemens and E.ON negotiated a unique energy performance contract that provided for two phases. In the first phase only a simple cycle plant was to be constructed. During a following 18-months testing phase, Siemens would be given full flexibility to test the new gas turbine. In this phase Siemens were accountable for the operation of the plant. Siemens paid for the gas supply and sold the generated electricity. The first gas turbine for the new power plant Irsching IV was delivered by Siemens in April 2007 and was first fired in December 2007 (see Fischer and Nag 2011). During the 18-month testing and validation phase, Siemens conducted several measurement campaigns. Under ISO conditions the gas turbine has a rating of 375 MW, but it was loaded to over 400 MW during open cycle testing. On August 28th, 2009 the testing and validation phase was completed and with this completion the second phase of the contract between Siemens and E.ON began, the expansion of the Irsching IV plant to a combined cycle power plant. Construction was finished in January 2011, and during the next 6 months the combined cycle was tested. In July 2011 the plant was turned over to E.ON.

All of the original development goals were reached or even exceeded. The combined cycle power of the 50 Hz model in Irsching IV reached 578 MW with an efficiency of 60.75%. The emissions of NOx were lower than 25 ppm and the CO emissions were below 10 ppm. Furthermore, due to the pure air-cooling system high operation flexibility was achieved. The fast loading mode allows a start-up time of 10 min to 350 MW. A ramp down to a minimum load at 100 MW or complete shutdown is possible in <30 min. After a total planning, development and testing phase of more than 10 years, the handover of the Irsching IV power plant to E.ON marked the start of the first commercial operation of the Siemens H-class Gas Turbine. The first commercial order came from Florida Power & Light (FPL) for six Siemens H-Class Gas Turbines to modernize the FPL power plants in Cape Canaveral and Riviera Beach (see Fischer and Nag 2011). A second order came from the South Korean client GS Electric Power & Services. A fully turnkey 410 MW combined power plant was built in Bugok. By January of 2017, Siemens had sold a total of 80 H-class gas turbines. After more than 200,000 total operating hours of the worldwide H-class turbine fleet one can attest that the original technical goals could be fully reached. The fleet has a reliability of over 99% and achieves an efficiency level of over 60% in combined cycle power plants. Especially the short startup times and the fast load-changing capabilities make the H-class gas turbine attractive for the market.

3.6 Summary and Discussion

Investment decisions in innovative research and development projects are accompanied by a high degree of uncertainty, different risk types, managerial flexibility during the project lifetime, and, finally, investment expenditures that are usually irreversible. These characteristics make project decisions in research and development very difficult. Very high investment expenditures and a frequently long time lag between the investment expenditures and the market launch underscore the need for a well-analyzed allocation of scarce resources to projects in research and development.

At the beginning of the last decade, Siemens was considering the development of a new type of gas turbine. The technological development promised to enable larger and more efficient gas turbines, and the competitors had already started to develop new gas turbines. Additionally, the development of a whole new generation of gas turbines might have enabled the integration of Westinghouse technologies into a common family of gas turbines after Siemens' take-over of Westinghouse some years previously. Several global trends in energy markets required new technological solutions to better meet the increased demands of customers. In many countries liberalization and privatization of energy markets had increased the price pressure on power plant operators. Thus, power plant operators asked for more cost-efficient solutions. Moreover, trends toward more decentralized energy generation and the increasing importance of renewably generated electricity was accelerating the demand for highly flexible turbines and power stations. To meet the market demand, Siemens decided to set ambitious goals for the new product family in an early phase of the development process. The main goals were efficiency rates above 60%, a very high degree of flexibility with short ramp-up times and very good partial load behavior, and a reduced life-cycle cost for clients by 7–8% over the life time of the gas turbines.

The global growth and market outlooks were promising. Worldwide power demand was growing at nearly 3% annually and the share of gas-based production had also increased in past and was forecast to grow further after the planned market launch of the new gas turbine series. However, Siemens carefully analyzed all relevant risks prior to the development decision and identified a whole range of uncertainties in this process. Market-related risks mainly stemmed from uncertainties about the gas price, one of the most important primary variables. Furthermore, the actions of competitors, the development of worldwide power generation and of alternative sources of energy, as well as regulatory developments meant more uncertainty about market success and potential sales.

With the exception of the market-related risks, no substantial technological risks were identified. The most problematic risk is that the development goals cannot be realized at all or at least not with the planned expenditures. A delayed market launch or even the abandonment of development may be the consequence. However, in addition technical problems can cause high costs after a market launch. If the turbines are not sufficiently tested, long downtimes can cause high warranty costs. To lower and manage the technological risks Siemens applied the Siemens Risk Assessment tool. All relevant failure risks were identified and quantified in terms of the failure probability and potential downtimes and warranty costs. The risks were updated after all relevant development stages when new information emerged.

Besides the various types and sources of risks that had to be considered and managed during the decision and development process, Siemens had to make several decisions after the initial project decision. Classical capital budgeting theory usually assumes only one initial investment expenditure. However, in a research and development project the investment expenditures usually occur at several different stages. It is not necessary to spend all funds once the project decision is made. Such a project structure allows management to maintain flexibility in the project and even halting it if new information makes this advisable. Siemens explicitly considered this possibility utilizing a stage-gate-process. Five development stages were defined when the project was started, and for each stage a gate was defined. At each gate new information should be taken into consideration. From stage to stage, continuously improved information about market attractiveness, market development, technological feasibility, and development expenses should enable better abandonment or continuation decisions. An explicit continuation decision is necessary to commence with the next stage and spend the related funds.

The project was initially started on October 1st, 2000. In March 2001 the product strategy was defined and the conceptual design phase was started after the market attractiveness of the planned gas turbine was confirmed. During the conceptual phase, the main technological decisions should be made. In addition to the managerial flexibility to abandon the project at a number of stages, Siemens enjoyed a lot of technological flexibility in the development process. During the early development phase, it became more and more obvious that higher operational flexibility of combined cycle power plants would be necessary in the future. One way to obtain higher operational flexibility was to use a pure air-cooling system instead of a steam-cooling system. Though this made achieving the ambitious efficiency goals more complicated, Siemens decided to use a completely air-cooled concept.

The development project took 7 years, and the first gas turbine was delivered in April 2007 for testing. To avoid high failure and warranty costs, Siemens decided to start an innovative cooperation project with the German power plant operator E.ON to test the newly developed gas turbine. In Irsching a new power plant was built to test and validate the first H-class Gas Turbine. In an 18-month testing phase, Siemens was solely responsible for the operation of the power plant and could test it in a real plant environment with a grid connection. After the successful testing phase, the power plant was extended to a combined cycle power plant. All development goals were reached or exceeded. In a world record run, it reached 578 MW with an efficiency of 60.75%. The combined cycle power plant was finally sold and has been operated by E.ON ever since. However, due to the comparably high gas prices and the increased share of renewable energies in Germany it now operates as a backup power plant to ensure a high level of grid stability. The takeover of the power plant by E.ON meant the completion of a more than 10 years development phase for the Siemens H-class Gas Turbine, which had sold 80 times by January 2017.

The case of the Siemens H-class Gas Turbine demonstrated how to successfully perform a complex development project in a highly uncertain environment. The development and the market launch was a great success for Siemens. Today, the competitors have also developed H-class gas turbines with efficiencies of more than 60%, but Siemens still holds a strong position in the market. To maintain its competitive advantage and to offer a superior product to its clients, Siemens will continue to further enhance the performance of the Siemens H-class Gas Turbine. Efficiency of more than 63% is currently targeted by Siemens and its competitors. Due to the very long product life cycles, the Siemens H-class Gas Turbine will remain a key product for Siemens in the gas turbine market.

References

Amram, M., & Kulatilaka, N. (1999). *Real options – Managing strategic investment in an uncertain world*. Boston, MA: Harvard Business School Press.

Black, F., & Scholes, M. (1973). The pricing of options and corporate liabilities. *Journal of Political Economy, 8*(3), 637–654.

Ceseña, E. M., Mutale, J., & Rivas-Dávalos, F. (2013). Real options theory applied to electricity generation projects: A review. *Renewable and Sustainable Energy Reviews, 19*, 573–581.

Cooper, R. G. (2002). *Top oder Flop in der Produktentwicklung*. Weinheim: Erfolgsstrategien: von der Idee zum Launch.

Cooper, R. G. (2008). Perspective: The stage-gate idea-to-launch process – Update, what's new, and NexGen systems. *Journal of Product Innovation Management, 25*(3), 213–232.

Dixit, A. K., & Pindyck, R. S. (1994). *Investment under uncertainty*. Princeton: Princeton University Press.

Fernandes, B., Cunha, J., & Ferreira, P. (2011). The use of real options approach in energy sector investments. *Renewable and Sustainable Energy Reviews, 15*(9), 4491–4497.

Fischer, W. (2011). Siemens H-class gas turbine, For pioneering efficiency with world class flexibility.

Fischer, W. J., & Nag, P. (2011, December 13–15). *H-class high performance siemens gas turbine sgt-8000h series*. Las Vegas, NV: Power-Gen International.

Friedl, G. (2001). *Sequentielle Investitionsentscheidungen unter Unsicherheit*. Berlin: Duncker und Humblot.

Friedl, G. (2003). Bewertung von Investitionen in die Entwicklung neuer Produkte mit Hilfe des Realoptionsansatzes. In U. Hommel et al. (Eds.), *Realoptionen in der Unternehmenspraxis. Wert schaffen durch Flexibilität* (Vol. 2, pp. 377–397). Berlin: Springer.

Granig, P. (2007). *Innovationsbewertung – Potentialprognose und-steuerung durch Ertrags- und Risikosimulation*. Wiesbaden: Gabler.

IHS Energy. (2016). *IHS long-term planning and energy scenarios*.

Merton, R. C. (1973). Theory of rational option pricing. *The Bell Journal of Economics and Management Science, 4*, 141–183.

Ratliff, P., Garbett, P., & Fischer, W. (2007). *SGT5-8000H Größerer Kundennutzen durch die neue Gasturbine von Siemens*. Sonderdruck aus FBG Power Tech.

Smith, J. E., & Nau, R. F. (1995). Valuing risky projects: Option pricing theory and decision analysis. *Management Science, 41*(5), 795–816.

Trigeorgis, L. (1996). *Real options – Managerial flexibility and strategy in resource allocation*. Cambridge, MA: MIT Press.

Trigeorgis, L., & Reuer, J. J. (2016). Real options theory in strategic management. *Strategic Management Journal, 38*, 42–63.

Valuing Product Lifecycle Management

Christopher Scheubel, Philipp Bierschneider, Holger Gierse, Ralf Hermann, and Michael Wokusch

In the first section of this chapter, we provide some background and essential information on product lifecycle management. First, we describe the concept of product lifecycle management (PLM) and analyze the influence of this concept on the value chain of product development. Subsequently, the potential value creation by means of product lifecycle management is presented, and the main value drivers of the concept are emphasized. In the next section, the cost related to implementing product lifecycle management is broken down. As this matter concerns IT investments, the concept of the Total Cost of Ownership (TCO) is introduced. The following section describes the benefits, costs and challenges which are related to the implementation of the Siemens PLM Software Suite®. It describes the general implementation process of adopting the Siemens PLM environment. Subsequently the findings are illustrated with the example of a Siemens business unit which implemented PLM throughout the lifecycle of their products. The case study illustrates our theoretical considerations with real business cases. The chapter concludes with a summary of findings and provides an outlook on future evolution in product lifecycle management.

C. Scheubel (✉)
TUM School of Management, TU München, München, Germany
e-mail: christopher.scheubel@tum.de

P. Bierschneider
DF FA SE DS DI, Siemens AG, Fürth, Germany

H. Gierse
DF CP S&MK BA, Siemens AG, Fürth, Germany

R. Hermann • M. Wokusch
GS IT DE PRG, Siemens AG, Fürth, Germany

G. Friedl, H.J. Kayser (eds.), *Valuing Corporate Innovation*, Management for Professionals, DOI 10.1007/978-3-319-64864-4_3

1 Characteristics of Product Lifecycle Management

This section first introduces the concept of product lifecycle management in the context of this work and provides a short historical outline of how the concept has evolved over time. Subsequently, the influence of product lifecycle management on the development value chain is outlined. These two subsections represent the basis for the analyses of the following sections which examine value creation and the costs associated with product lifecycle management.

1.1 Definition and History of Product Lifecycle Management

The central purpose of every company must be to generate value and to improve its current competitive position. Especially with regard to the industry sector, these objectives are closely related to developing, producing, and selling innovative products. Along with the omnipresent advance of information technology (IT), novel capabilities of software based product development and manufacturing are set into motion. These support programs integrate all necessary steps within the lifecycle of a product and thus generate a holistic approach to new development processes. This enables a company to create products with an increased product-to-market fit, improved quality and a shorter time-to-market period. This approach, consequently, results in an improved competitive positioning for the company. Despite the huge advantages of product lifecycle management (PLM) over traditional processes, holistic functioning support systems are still not commonly implemented in industry (see Eigner and Stelzer 2009, pp. 5–6). In order to improve this situation, an analysis of the value drivers and the related implementation costs of such software-based development tools are absolutely necessary.

Product lifecycle management (PLM) consists of product as well as production development. It is a concept that facilitates the requirements- and manufacturing-planning processes of new products with the support of software tools. Design, definition, engineering, production planning, and the information related to these stages are integrated and digitally processed (see Stark 2015, pp. 1–29). Efficiency is improved as there is a common data backbone, which facilitates easy communication between departments globally. 3D engineering of products and digital exploded assembly drawings enhance the creative work of manufacturing and production engineers. All data created during the design and engineering process as well as the requirements, supplier-/pre- and post-sales documentation and simulation data can subsequently be transferred to manufacturing and support services. This also applies to any product changes and amendments which have to be made during the product's lifecycle. Consequently, implementing PLM ensures the seamless integration of all departments involved in the development and manufacturing processes of new products. Even customers or suppliers can be involved which supports an improved product-to-market fit as well as lean production processes.

Historically, the year 1963 can be regarded as the genesis of computer aided design. In their seminal work, Ross and Rodriguez (1963) describe how computer aided design (CAD) systems could function in an industrial environment. The authors also outline CAD's potential use in product development processes (see Ross and Rodriguez 1963, pp. 306–308). Early work on design support systems was predominantly driven by Massachusetts Institute of Technology (MIT) research. The first industrial application of a computer aided product development system can also be dated back to 1963, when the computer aided manufacturing (CAM) system "DAC 1" was introduced by General Motors (see Myers 1998, p. 49). During the late 1960s and the 1970s, the development of real-time 3D raster graphics made more advanced design applications possible (see Myers 1998, p. 50).

It was not until the 1990s, however, that IT began reshaping product development on a broad industrial scale (see Nambisan 2003, p. 6). In the years after the turn of the millennium, design and engineering technology matured and IT tools were used to support several steps of product development and manufacturing. In the years after 2010, initial attempts were made to integrate the whole product lifecycle with one suite of software (see Srinivasan 2011, p. 464). This development towards a holistic development software had three predominant drivers: there were standardized product data models available, data sharing concepts emerged, and robust middleware facilitated the implementation of software modules (see Srinivasan 2011, p. 464). This integration of development steps led to the new concept called PLM, which effectively covered the whole ideation, design, engineering, production and after sales process of a product. This concept is represented by the Siemens PLM Software Suite®, which is one practical realization of this concept. Parallel to the software development, progress in theory and practice concerning the process-oriented organization was also made. Product development processes became increasingly market orientated. The emergence of a holistic PLM view can thus be seen as the result of two developments, process-oriented organization and enabling software tools, the evolution of which mutually enhanced each other.

Initially, PLM started out as the integration of engineering information systems, mainly CAD, computer aided engineering (CAE), computer aided manufacturing (CAM) and product data management (PDM). CAD systems are used for 3D representation and design purposes. An engineer or designer can analyze the shape of products and also simulate the interaction of several design elements using these tools. When the durability of components and material characteristics are an additional concern, a more advanced method of computer based support is necessary. With the help of CAE, the dimensioning of products and components can be adjusted to the specific application for which they are being developed. To test the durability of these products, the method of finite element analysis (FEA) is generally applied. Additional highly specialized tools for thermal or electric simulation are applied which can also be integrated into the PLM data backbone. FEA is a numerical approach to solve partial differential equations which are necessary for simulating the performance of solid body parts, a method which has become essential, e.g. for the engineering of engines, turbines, and parts of the car-body and chassis.

Once the computer aided engineering process is finished, a prototype is usually made. This can be done virtually with the virtual prototyping (VP) technique before costly physical prototypes are built. The virtual prototype can be displayed for the purpose of design judgment, and simulations can be carried out to ensure the functionality of the product and the interactions between different product components. During this phase the manufacturability and serviceability of the product are also verified. In effect, the adequacy of production with the desired manufacturing equipment as well as the access to components for service purposes are reviewed. Normally, the interaction between engineering and prototyping is iterative in nature. Several loops are necessary before the finished digital design and production blueprints are transferred to the manufacturing department. In some cases, the prototyping process is already performed in the manufacturing department; consequently, the expertise in manufacturability of a product directly enters the prototyping process.

CAM is the logical extension of CAE and the prototyping process. As the data of the product is already digitally available, the next step is to integrate this data directly into the production process. Computerized numerical control (CNC) programs operate the machine tools for the newly developed products and also determine the size and timing of production batches. With CAM software, the development time of CNC programs can be minimized, as data generated by CAM and CAE processes can be seamlessly converted with post processors and transferred to the production machines. It is also possible to integrate the existing design, engineering and manufacturing data into operation maintenance services, thus facilitating improved utilization. The mean-time-to-repair can be significantly reduced, though an entirely integrated end-to-end product development is still rarely found in practice.

CAD, CAE, CAM, and the PDM backbone are the core components of PLM. This concept also offers novel methods for integrating other parts of the product lifecycle, e.g. the supply of raw materials and components as well as customization during the production process. The involvement of customers in the design and engineering process leads to an improved product-to-market fit and the possibility of manufacturing highly customized products. Naturally, the parties involved have to use the same product development and data management platforms, wherever possible, to ensure seamless interoperability. Current developments of PLM include the integration of big data analytics (see Li et al. 2015, pp. 667–684), the orientation on sustainability driven development (see Vila et al. 2015, pp. 585–592), and the compatibility with product lifecycle services (PLS) (see Wiesner et al. 2015, pp. 36–41).

1.2 How Product Lifecycle Management Is Transforming the Whole Product Development Process

In order to understand the influence of product lifecycle management on the product development process, we need to consider the individual steps necessary for product development. Figure 1 shows the necessary core steps of product

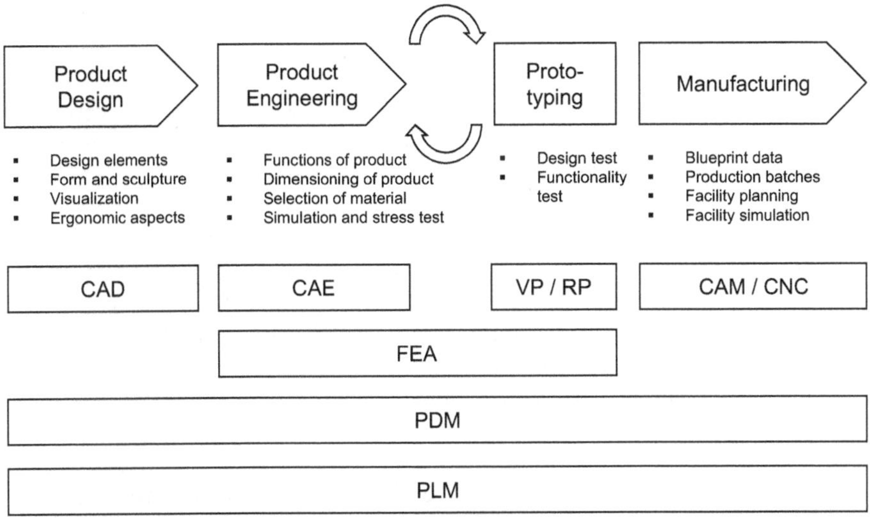

Fig. 1 Core steps of the product development process with the respective software based support tools (Source: Own illustration based on Chandrasegaran et al. 1991, p. 214; Eigner and Stelzer 2009, p. 6)

development and the respective supporting software. The process can be extended to begin at ideation and end at product disposal. However, we shall concentrate on the core product development steps of design, engineering, prototyping, and manufacturing.

Product design comprises creative work on the form, appearance, and functionality of a product and should also incorporate ergonomic as well as technical constraints of the product. CAD tools are employed for these design processes, as they are able to visualize the new product digitally, which makes creative comments and approval processes easier. The data created during this process are automatically stored in the PDM of the PLM system. The subsequent step enriches the CAD data with engineering information. Here the product material has to be chosen and specifics concerning the manufacturability are elaborated.

Simulations and FEA test the durability of the product, while its interaction with other components can be examined using CAE software. Data and information enters the PDM system which serves as a basis for prototype creation. With VP tools prototypes can be displayed for design, assembly, and functionality tests. In this way, costs and time for the production of physical prototypes can be saved. Iterations between CAE and Prototyping have to be continued until the final design is approved and the manufacturing process can begin. Once the design and engineering of the product has been finished and approved, the data are automatically transferred to the CAM tools and at the same time translated to CNC code, which operates the manufacturing machine tools. This automated transfer of data also enables an automatic design change process. Any changes made in the product design are instantly implemented in the manufacturing process. All 3D engineering data and supplementary information are continuously supplied to the PLM backbone for the purposes of operation and maintenance services.

By implementing these software-based tools, every development step becomes standardized and traceable, which improves data and process quality. Moreover, an overarching PLM system will enable comprehensive collaboration including release processes and workflows, which can be a key to outsourcing tasks globally.

Following the implementation of a PLM environment, employees might feel that there has been a loss of individual freedom, a factor that at first might outweigh the considerable advantages offered, such as transparency and process discipline. Consequently, along with the implementation of a new IT infrastructure, the organizational processes and a mindset change of the workforce is necessary. This facilitates efficient collaboration and ensures adequate conditions for a successful and frictionless IT implementation.

2 Potential Value Creation of Software Based Product Lifecycle Management

In this section we will analyze how value creation in the product development process can be generated by product lifecycle management. First, the potential value creation resulting from implementing a PLM environment is shown. Time-to-market, as well as cost and quality effects are described. Subsequently value levers based on a product's lifecycle are analyzed. We will show that PLM can have a positive influence on revenues while decreasing the development cost or costs in subsequent processes.

2.1 Empirical Results on Value Levers of a Product Lifecycle Management Environment

Durmuşoğlu and Barczak (2011) empirically analyze which IT tools have positive effects on the success of newly developed products. The authors divide the product development process into three stages, the discovery, the development and the commercialization stages, in order to analyze the impact of eleven commonly used IT tools on the success of these phases. The authors employ a sample of 212 US and Canada based companies, which includes various manufacturing branches as well as services like finance and insurance companies. The new products were then classified by three categories: new product market performance, new product innovativeness, and new product quality. The market performance is measured on a scale from "below expectations" to "above expectations". The innovativeness was classified by uniqueness concerning technology, features and benefits compared to other products on the market, while quality was measured by strong agreement or disagreement concerning the customer expectations compared to competitor products. The results were significantly positive for the tools of decision support software across all development stages; for file transfer protocols across all stages, for secondary data in the discovery stage, for online needs surveys in the commercialization stage, for virtual prototyping in the discovery stage, and for concept testing software in the discovery as well as the development stage (see Durmuşoğlu and Barczak 2011, pp. 322–326).

Research suggests that the success of new products on the market can be significantly increased when IT is used during the development process. In a 2003 survey of the Product Development and Management Association (PDMA), the percentage of successful products was about 60%. Since 1995, this success percentage has not changed greatly (see Barczak et al. 2009, p. 6). Nambisan (2003) proposes that during the 1990s IT reshaped the entire development and innovation process of industrial companies (see Nambisan 2003, pp. 7–8). The author focuses on the steps of process management, project management, knowledge management, and collaboration/communication (see Nambisan 2003, pp. 6–8). Various industries have picked up on this trend and increasingly use software-based tools to guide their product development processes. According to a survey by McManus (2005), nearly 80% of a typical engineering job within the aerospace industry is non-value creating (see McManus 2005, pp. 14–15). This concerns the time a project is not being worked on or non-value creating tasks are being performed, for example, if data is passed to another department which is currently occupied with other projects or when an engineer has to build a new file structure for the project data. IT tools supporting the product development process are intended to eliminate this lost time.

All of these analyses show that software based development processes are more successful in the categories of market performance, innovativeness, and product quality compared to their traditional counterparts. Specific development tools are also bound to certain stages in the development of products, a result supported by the growing number of firms using software based tools to guide their development processes. This analysis is based on isolated IT tools, however. If the whole process chain for new product development is considered, three predominant levers can be identified.

As depicted in Table 1, these levers can be categorized by time, cost, and quality. The positive effect of these levers concerns both outcomes of a well-organized industrial company: lean processes and the developed product.

Production processes should be lean, which means information as well as material should be available for the subsequent process step as soon as it is needed. The principle also implies that there should be as little material as possible in stock to decrease the cost of capital for inventory. Information should not be delayed to ensure short development and production times. Another aspect of lean processes is the elimination of waste in any form and consistent continuous improvement. There

Table 1 Levers of product lifecycle management which create value concerning industrial processes and products

	Time	Cost	Quality
Process	Lean-production	Stay-below-budget	Quality-compliance
Product	Shorter-time-to-market	Lower-development-cost	Exceed-customer-expectations

should be no necessary recalculation due to malfunctions in the process planning procedures. All processes should be done "right the first time". The quality aspect within the production process connects compliance with certain product-standards: The production process has to ensure that all standards necessary to effectively market the product are fulfilled. Meaning to exactly match the customer demand and neither under- nor over-fulfill requirements. The quality standards of the product must be adhered to, while the incurred costs must be restricted.

With regard to the product, a shorter time-to-market implies that the time period from ideation to the market launch of a product has to be shortened, which will lead to competitive advantages for the company in the market. A PLM environment will lead to lower costs per product launch due to reduced working time for engineers and designers when software based support tools are employed. The centralized data backbone enables the company to optimize R&D as well as production resource allocation. The whole value chain can be distributed globally. Production costs can be saved due to optimized manufacturability and a faster ramp-up of production. The quality aspect of the product is focused on customer expectations. PLM enables early customer integration into the development process and makes the conversion of customer requirements transparent. Thus an enhanced adherence to customer expectations can be achieved. When an OEM directly integrates his suppliers into the development processes, the innovation cycles will be shortened. This reduces the time-to-market while improving supplier relations.

Figure 2 illustrates how a shorter time-to-market can be realized by implementing a PLM environment. In traditional product development, specialization of single processes led to a consecutive rather than a simultaneous workflow.

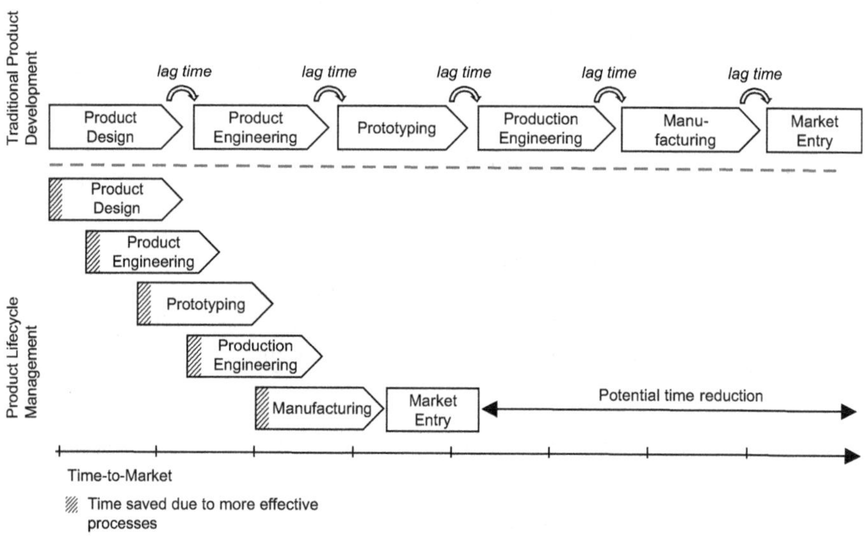

Fig. 2 Realization of a shorter time-to-market by implementing PLM (Source: Own illustration)

Every step had to be finished completely and approved before the next step in the development process could start. The implementation of PLM, however, allows for an integrated product development and simultaneous engineering. The concept of integrated product development combines organizational, methodological, and technical capabilities to enable a holistic approach to product development. The interdisciplinary collaboration and parallel work of product development, production, and sales can be described as concurrent engineering (see Ehrlenspiel and Meerkamm 2013, pp. 193–210 and 227–229).

As depicted in Fig. 2, the traditional product development process took several steps. Depending on the particular industry, the product had to be designed, engineered, prototypes had to be made, and the product had to be manufactured before it could finally be launched on the market. These steps had to be performed consecutively, as the information generated during these steps had to be completed and approved, before being passed on to the next step. Time lags were evident between the steps as handovers could not be synchronized between the departments involved.

Within a PLM environment, a common transparent data backbone is combined with continuous approval processes, which enable simultaneous work on the same product in different departments. The software directly informs subsequent working steps when information is partly ready and approved for being passed on. Consequently, there are no time lags as information transfers are synchronized automatically.

2.2 Identifying Product Lifecycle Management Value Levers and Using the Product Lifecycle Concept

The product lifecycle is a concept that shows the profits made with one product line during all phases of its life. It resembles a wave with two inflexion points and represents five stages of the product's lifecycle: incubation, growth, maturity, decline and end of life. At the beginning losses are made, as there is a cost related to product development and the ramp-up of production. Subsequently, the product sales period starts at the first inflexion point of the curve. Profits are made when the sales revenues of the product outweigh the initial cost of development and production ramp-up. The profit period lasts until the product is removed from the company's product portfolio. The cumulated total profits are equal to the integral of the entire product lifecycle curve. To maintain its business, a company will need a new product before the end of the initial product's lifecycle.

The levers obtained through the implementation of a PLM environment on the basis of a product's lifecycle are shown with arrows in Fig. 3. While the grey dashed line indicates profits made without PLM, the solid line illustrates the achievable profit development with a product lifecycle management system. Four key levers maximize the profit integral under the curve.

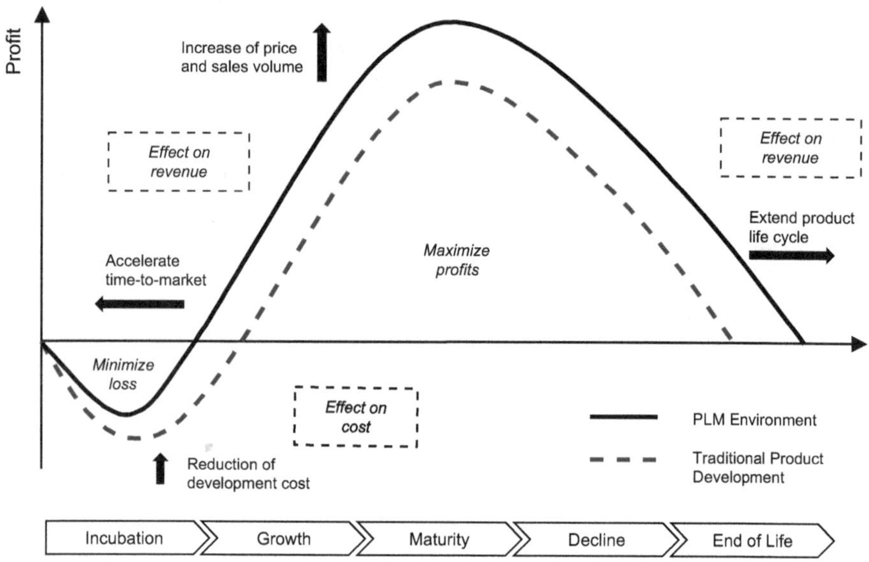

Fig. 3 Product lifecycle with levers made possible by a PLM implementation (Source: Siemens company data)

The shorter development times and the reuse of engineering data reduce the development cost of the product. An accelerated launch enables the company to charge higher prices, as they can capitalize on the competitive advantage of an early market entry, which will shift the profit curve upward. A PLM environment also provides a company with the ability to react faster to market conditions. Customer requirements can easily be processed to supplementary products or product variances. This shorter response time leads to faster product modifications which can extend the lifecycle of the product so that the period in which profits are made can be increased. The value levers shown above are thus: a lower development cost, higher prices in the market, an extension of the product's life time and a faster time-to-market (Fig. 3).

Another cost effect worth mentioning when analyzing a product's lifecycle is based on the fact that early ("frontloaded") effort in product development will lead to lower costs during manufacturing (Fig. 4). The cost effect is tremendous, as the cost of modifying products during later development stages or even during the manufacturing process increases exponentially. For example, the costs for realizing product changes rise significantly when the decision to change a product feature is made later in the development process as all previous documentation and designs have to be changed as well. In the worst case, specially-produced machine tools have to be altered or re-built, which is why most development effort should be dedicated to early product feature definition and product manufacturability. The aim of a PLM environment is to avoid changes in the later development stages.

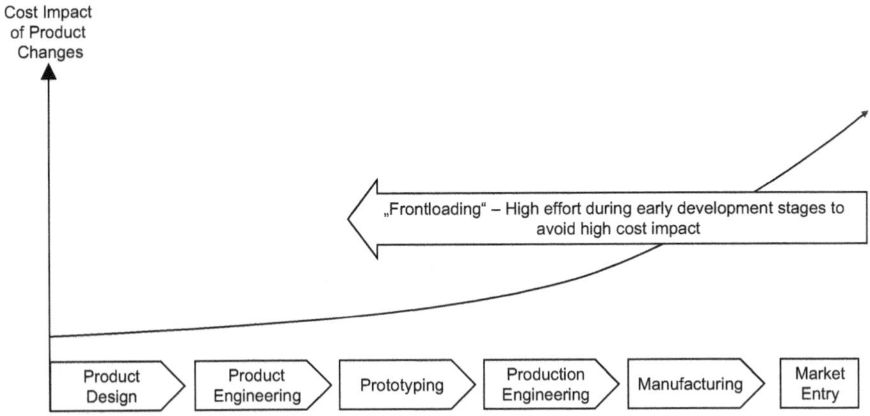

Fig. 4 Cost impact of product changes along the development stages (Source: Own illustration)

3 Total Cost of Ownership for Software Based Product Lifecycle Management

In this section we first describe how the Total Cost of Ownership concept evolved over time. We outline the reasons behind the development of such a cost accounting framework, and subsequently demonstrate how the Total Cost of Ownership concept should be applied in the context of PLM.

3.1 Literature on the Total Cost of Ownership Concept

Corporations in competitive environments and under margin pressure have to closely evaluate the economics of IT investments. The Total Cost of Ownership (TCO) is a concept which incorporates all direct and indirect costs of an IT system. It is commonly used by corporations to determine their IT spending. These expenses made up roughly 3.4%[1] of corporate revenues worldwide in 2009 (see Potter et al. 2011, p. 1). TCO yields a more accurate estimation of the costs than traditional total cost accounting or capital expenditure measures, which were formerly applied as the basis for strategic decision making, especially concerning IT investments (see Ellram and Sue 1998, p. 55). Accordingly, the TCO of the Siemens PLM Software Suite® has to be considered thoroughly when planning such an investment. The value generated by the software implementation must be compared to its TCO (see Geißdörfer 2009, pp. 694 and 705–706).

[1]These figures are based on a Gartner Inc. survey of 1756 companies across 22 industry sectors.

The genesis of the TCO concept can be traced back to 1987 when Gartner Inc., a company which specializes in information technology research, announced that IT related costs were quite obscure and investment decisions inaccurate due to overvaluing initial costs. Zarnekow et al. (2004) argue that, besides taking the initial investment into account, IT managers also intuitively consider the operation and maintenance cost of their acquisitions. However, there is no standardized process of accounting for these costs (see Wild and Herges 1998, p. 3). Their survey of 30 IT applications in three companies shows that the share of cost during the operation phase is 79% of total costs on average (see Zarnekow et al. 2004, p. 185), which means the mere consideration of the initial investment costs of a software application may lead to highly biased decisions. The concept of TCO was mainly applied in the context of IT expenditures with similar structures, such as workplace PCs, LAN networks or server infrastructures (see Wild and Herges 1998, pp. 3–4), as traditional concepts failed to account for post-acquisition costs. TCO thus became increasingly important as information technology became an important productivity driver for companies. Aspects which have to be considered when evaluation IT investments include (see Wild and Herges 1998, p. 6):

- The indirect cost which occurs, for example, when colleagues support each other when new software is implemented.
- The transparency of IT costs, which did not exist and made benchmarking comparisons between companies nearly impossible.
- Inaccurate estimations of the impact of investment decisions. Management was confronted with minimum cost estimations which biased the decision-making process on investments to be made.
- The unsatisfactory results of all IT related data, as there was no standardized system for recording the necessary data. Furthermore, the cost related to recording the data was overestimated.

Concepts similar to TCO include total cost accounting, product lifecycle costs, and lifecycle costing. Similarly, the TCO concept encourages purchasing managers to take a long term perspective on their buying decisions. These concepts aim to take into account all costs related to purchasing decisions as well as the cost and business impact that the purchase has on other departments (see Ferrin and Plank 2002, p. 1). The concept of the Total Cost of Ownership goes beyond the cost of manufacturing goods. It includes all licenses, service contracts, and capital costs. As with software products, the licenses, services and maintenance represent a significant component of the total costs which the company has to bear.

Gartner Inc.'s first Total Cost of Ownership model from 1987 focused on desktop PCs, since they realized that traditional cost accounting concepts would not correctly reflect a company's IT related expenditures (see Wild and Herges 1998, p. 3). The recent development concerning cloud-based architectures, e.g. Microsoft Office 365, has raised the percentage of license and service fees even further over the initial investments into hardware and software. Today, TCO is the commonly applied framework to assess IT investments.

The Gartner Inc. model divides IT related costs into direct and indirect costs. Direct costs, depreciation of hardware or the salaries of IT personnel, can be assigned to the IT department. Indirect costs are not so easily traceable, as they are caused by efficiency losses connected to the use of an IT infrastructure. Engineers, for example, cannot do their jobs, when the application server of the product design tool is not available. This type of cost is very hard to trace, as there are no directly related receipts or invoices (see Wild and Herges 1998, pp. 10–11). These costs have to be included, though, as otherwise the cost of any IT project or implementation would be underestimated. Meckbach (1998) confirms the difficulty of correctly applying the TCO concept, as there are constituent elements which are especially hard to measure, for example user satisfaction related to the purchase of new hardware or the productivity increase related to a software update.

It should be noted, however, that direct and indirect costs are inversely related to each other. When the company saves on direct costs, e.g. personnel in IT support, then indirect costs will automatically be higher. Employees will spend more time getting used to the new software features and will also increasingly help each other with peer-to-peer support, which instantly has negative effects on the productivity of these employees and increases the indirect costs. On the other hand, the direct costs may also be too high when there is too much IT support and administrative IT resources are not used efficiently (see Elsener 2005, p. 218). Figure 5 illustrates this inverse relation and shows that the equilibrium is located at the intersection of the curves, meaning that direct costs should not be saved on too much, though excessive spending will naturally have a negative cost effect as well. As the minimum cost can rarely be achieved, a range of acceptable costs is indicated.

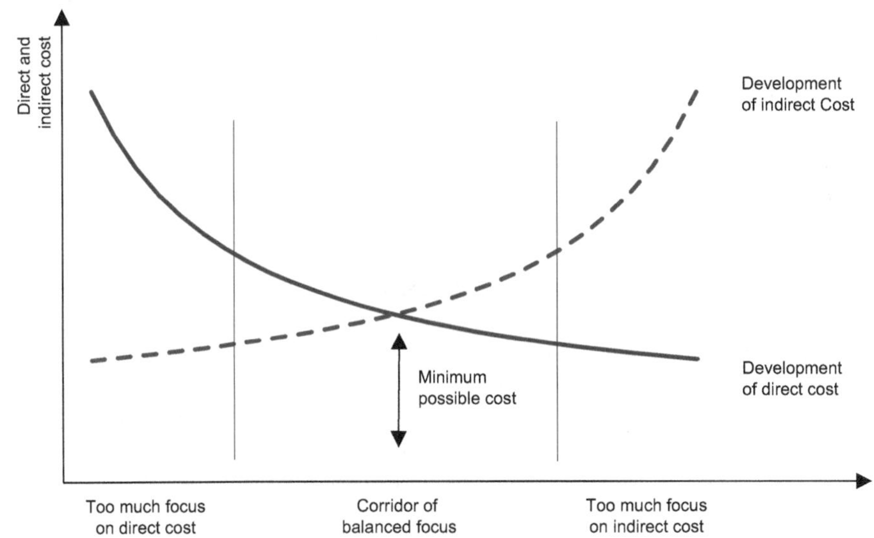

Fig. 5 The interaction between direct and indirect costs (Source: Illustration based on Elsener 2005, p. 218)

3.2 Applying the Total Cost of Ownership Concept to Product Lifecycle Management Software Solutions

As related in the previous section, the TCO model originally developed by Gartner Inc. 1987 contained a comprehensive set of direct as well as indirect costs. Direct costs include all costs which are incurred due to IT department activities, whereas indirect costs result from inefficiencies when using the IT infrastructure. The most important subsets of these two categories is listed below (see Elsener 2005, p. 208).

- Direct costs: Depreciation of hard- and software, licensing fees, software updates, replacement parts, salaries for IT-personnel, IT-infrastructure operation, network services and administration.
- Indirect costs: Peer-to-peer support, downtimes, training seminars, and user developed software.

Direct costs are mainly based on hardware and software utilization. Depreciation of hard- and software represents an accounting based cost distribution of the initial investment over the first 3–6 years of utilization. Depreciation can only be applied, when there is an asset on the balance sheet, though this is not the case with software or hardware licenses. Updates are software improvements which the company normally has to pay for. The cost item of replacement parts includes only hardware components, while moving parts, like hard drives and the cooling system, have to be replaced on a regular basis. Salaries for IT personnel are the gross earnings, workplace equipment, health insurance and other social benefits. Additionally, there are fees to be paid for external IT resources. Operation of IT infrastructure generally includes the electricity necessary for the IT equipment, and costs for network services like the fees for data transfer. This cost can vary depending on the transfer speed necessary for the company. The administrative cost includes the installation of the IT infrastructure and support activities during operation. Naturally, only those activities are accounted for that are not handled by proprietary IT personnel. Administrative costs occur due to the setup process and operation of the IT department itself, but this includes only services they provide for themselves and not those for other parts of the company.

Indirect costs are based on efficiency losses caused by the implementation of new IT solutions. When a new software is implemented, employees have to be trained to work with it. Thus, there is a cost for the lost productivity as well as for the training classes. This efficiency loss can be mended by the organizational change process, which has to accompany and support the IT implementation. The effects of organizational change management will be explained in Sect. 4.2. During the start-up phase peer-to-peer support plays an important role. Employees will help and support each other when learning to work with the new software, which also causes efficiency losses. Downtimes of a new software are yet another cause for lost working time. The more people are dependent on the software, the higher the downtime-related cost is. Small user-developed scripts or macros improve efficiency when working with the new program; however, the coding and development time of these small programs consumes time resources (see Elsener 2005, pp. 211–218).

TCO takes all of these direct and indirect costs into account and thus offers a holistic view of an enterprise's IT related expenditures. The cost factors, as depicted in Fig. 5, are highly interrelated. If, for example, hardware costs are saved by purchasing medium quality computers, the operation and maintenance costs will be higher. Due to repair times the lost employee productivity will also increase. Reluctance to invest in centralization or standardization will reduce the effectiveness of IT performance. The necessity for extensive communication causes additional costs when no centralized information backbone exists (see David et al. 2002, p. 3). Zarnekow et al. (2004) confirm this interdependency, showing in their survey of 30 corporate IT applications that low initial investments are related to higher subsequent costs during operation—and vice versa (see Zarnekow et al. 2004, p. 185). Martens et al. (2012) apply the TCO concept to cloud computing; the following table is based on their findings. The depicted scheme can be transferred to the case of the Siemens PLM Software Solutions®.

Table 2 lists more cost items than found in Gartner Inc.'s core TCO concept. The application of the TCO framework or the framework shown in the table above is a complex task, however. Ferrin and Plank (2002) come to the conclusion that no

Table 2 Total cost of ownership concept for cloud computing (Source: Martens et al. 2012, p. 1568)

	Cost type	Cost factors
1	Strategic decision making to initiate the IT implementation project	Work time, consulting services, information to support the decision making process
2	Evaluation and selection of service provider	Work time, consulting services, information to support the decision making process
3	Service charge IaaS[a]/Hardware cost	Computing power, storage capacity, inbound data transfer, outbound data transfer, number of queries, domain, SSL certificate, license, basic service charge
4	Service charge PaaS[b]/Hardware cost	User-dependent basic charges, storage capacity, inbound data transfer, outbound data transfer, provider internal data transfer, extra user data storage capacity, extra user document storage capacity, queries to the Application Programming Interface, sent emails, database, secured logins, connections with other providers' applications
5	Service Charge SaaS[c]/license cost	Access to the service system, charge per user
6	Implementation, Configuration, Integration and Migration	Work time, porting process, efficiency cost, downtime cost, consulting services
7	Support	Work time, support costs, problem solving
8	Initial and permanent training	Preparation time of internal employees, participating time of internal employees, instruction material (mat), external consulting services
9	Maintenance and modification	Work time, service fees, support costs
10	System failure	Loss per period of time
11	Backsourcing or discarding	Work time, porting process

[a]Infrastructure as a service
[b]Platform as a service
[c]Software as a service

standardized model exists for calculating the TCO. There are a large number of possible cost drivers which can be considered; consequently, the company's specific situation and the IT project's circumstances have to be analyzed before selecting an appropriate subset of cost drivers (see Ferrin and Plank 2002, p. 16). As the findings of Martens et al. (2012) are based on the recent development of cloud computing, their list of drivers is applied in the case study examined in the following section. A possible subset of the IT acquisition of a comprehensive PLM environment is given below and will be applied to the case study in the subsequent section.

4 Case Study: Implementing the Siemens PLM Software Suite®

This section illustrates and analyzes an implementation process of a PLM environment at Siemens. Primarily the capabilities of the Siemens PLM Software Suite® are outlined and the challenges of an implementation as well as the related benefits are described. Subsequently, the implementation process is illustrated with the example of a Siemens business unit. Finally, we analyze the costs as well as the benefits resulting from the implementation and summarize our findings.

4.1 The Capabilities of the Siemens PLM Software Suite®

The Siemens PLM Software Suite® is a modular software package which integrates supportive development tools along the product development cycle. It is designed to conform to the principle of the possibility to design and produce anywhere in the world. It enables global collaboration along the whole process of product development. The software has an open architecture, mainly for evolutionary reasons, as it has to be enhanced continuously to follow market and customer demands. New parts of the software had to be continuously integrated, which made an open software architecture indispensable. The application of the software enables the headquarters to coordinate, monitor and control the whole product development process.

Table 3 segments the constituents of the Siemens PLM Software Suite®. Every subsystem is integrated into the common data backbone which enables several departments and people to work on the same project at the same time. It ensures that everybody is on the same level of progress and works with the same constantly updated data. The NX® toolsets enable product design, simulation and the planning of efficient production processes. Tecnomatix® is a separate tool which can be used to plan plant and factory layouts with the ability to simulate material flows and production processes.

Siemens Teamcenter® is a backbone system, which acts as a communication hub between the singular tools of the Siemens PLM Software Suite®. It is like an operating system, which provides a basis for applications to co-function with each

Table 3 The Siemens PLM Software Suite® Product Portfolio

Segment	Product	Application
Product engineering	NX for Design® NX for Simulation®	Faster design, simulation and production of superior products
Manufacturing engineering	NX for Manufacturing® Tecnomatix®	Plan the manufacturing process more efficiently
Lifecycle collaboration	Teamcenter®	Intelligent collaboration for more innovations
Mainstream engineering	Solid Edge®, Femap®, CAM Express®	Faster product design to shorten time-to-market
Specialized engineering	Fibersim®, Syncrofit®, SDE®, QPE®	Solution for complex development tasks
Simulation and test solutions	LMS Imagine.Lab®, LMS Virtual. Lab®, LMS Test.Lab®	Simulation and test environment for zero defect development processes

other. It also serves as a knowledge and information backbone that all tools communicate with. The Siemens PLM Software Suite® further offers tools for mainstream as well as specialized engineering purposes. Mainstream engineering comprises component construction tasks like the 3D modelling of gearboxes, while specialized engineering focuses on tasks such as optimizing the stability of carbon fiber structures. The virtual labs for testing and simulation reduce the efforts needed for costly physical prototypes and enable the production of highest quality products and components.

Figure 6 shows the Product Development Lifecycle of the Siemens PLM Software Suite®. As Siemens Teamcenter® is the common backbone for integrating several PLM tools, it shall be explained in more detail. Within Teamcenter®, processes are coordinated by an integrated workflow machine. Data compatibility is ensured by clearly defined release cycles which enables collaboration across departments, time zones, people as well as cultures.

As soon as a marketable product idea is developed, it can enter the cycle at the first step—the idea management tool. After that the requirements engineering follows, determining which features of the product are required by the market. The next step is the portfolio management which relates the product to the portfolio strategy of the entire company. It also indicates whether there are potential synergies in sourcing, manufacturing, marketing as well as cross-selling possibilities. In the project management tool, an entity is set up to guide the whole process from the engineering steps to the manufacturing processes.

The following steps are focused on design and engineering, starting with the electronics computer aided design (ECAD), going on with the design of the software (SW), the mechanical computer aided design (MCAD) and finally with the computer aided engineering (CAE). With ECAD, electronic circuit boards and integrated electric circuits can be virtually designed. The design of the SW concerns programming activities which enable the product's operation. MCAD is the design step which determines the shape and the mechanical interaction of product components. CAE comprises tools which are used for engineering analysis tasks

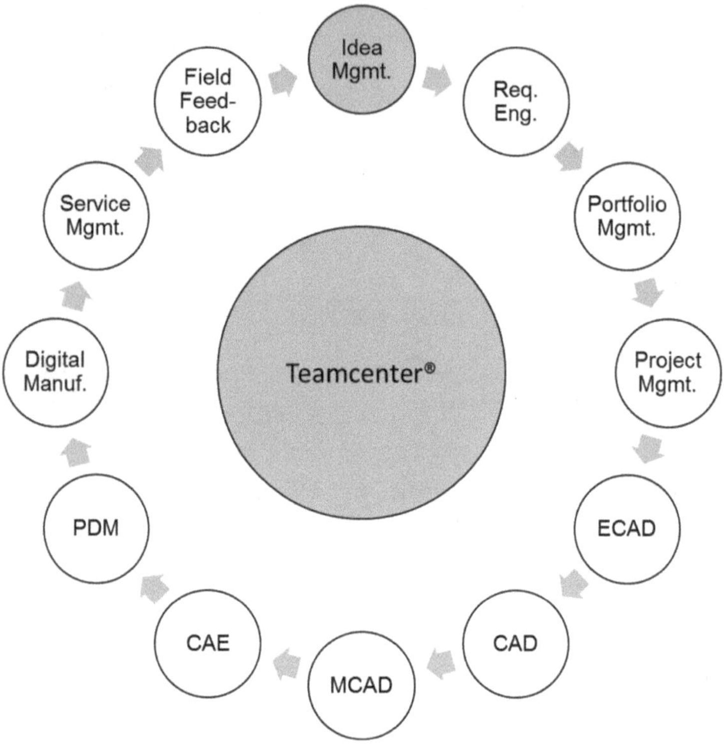

Fig. 6 Product development lifecycle of the Siemens PLM Software Suite® and its constituents (Source: Siemens company data)

like finite elements analysis (FEA), computer fluid dynamics (CFD) and multibody dynamics (MBD). The constraint conditions of an efficient manufacturing process have to be defined at this stage already.

Based on numerical methods and information technology, all design processes and engineering work is done with computer support. The advantages of the PLM software suite are compelling: This software can instantly identify whether the design matches the mechanical outline of the product, if the electrical engineering is appropriate and if the software suits the newly developed product.

The entire body of information flows into the product data management (PDM). All data concerning the previous steps is aggregated here. Thus, important information, for example, the exact requirements formulated by the marketing department or the product manager, is available for engineering when needed. The next step is the digital production planning and manufacturing process. During production planning a process sheet is created with all necessary manufacturing steps. For the digital manufacturing process all data concerning the product is directly fed into the machine tools. Before the implementation of the Siemens PLM Software Suite®, the data on design and engineering had to be translated and implemented into the computerized numerical control (CNC) systems of the machine tools. Now transference is possible automatically. The next two steps are service management and field feedback. The serviceability of products adds considerable value for the customer. The Siemens PLM Software Suite® offers several tools to facilitate the provision of maintenance as well as quick replacement of spare parts. Field feedback can enter the life cycle process at any time. All deficiencies are tracked and instantly result in a new requirement for the product. The new requirement triggers a new or amended feature which is then designed and constructed. After that the new and improved version of the product enters the manufacturing process.

The whole process of product development is shown in Fig. 6. It can be controlled, tracked and monitored by the Siemens PLM Software Suite®. It enables seamless collaboration between different departments and thus shortens the time to modify or improve as well as to market a product.

The Siemens PLM Software Suite® functions can also be taken into consideration along the value chain of product development. The whole concept and structure of the software suite is based on information enrichment and thus generates a forward oriented process with automated communication and few iterations between the departments involved. The data objects within the Siemens PLM Software Suite® environment represent ideation, requirements and features, project management, product data, quality and defect tracking, resource data, processes and plant data. These objects and their related documents are created and enriched during the product development process.

The whole development process is supported by the data backbone, which every process stakeholder can access. The first created data items are based on product ideas which are stored within pools. These ideas can come from customer requests or from upcoming general trends and innovations on the market. After the process of ideation the data of the generated ideas enter the requirement management. Here the ideas are formatted into to specific requirements and features that provide the product's characteristics and functionalities. The requirements and subsequently the features of the product are added to the documentation on the generated ideas. It is important that all information on the product is completely documented, so that, for example, an engineer can eventually cover several product requirements by one feature. Without a common data backbone, he would have needed to go to the respective departments and ask for the documentation about the ideas and requirements they initially thought of. Requirements and features are defined as well as the project management documentation set up. All relevant data for the project is collected here, including budget and resource planning which makes the first interaction with the corporation's enterprise resource planning (ERP) system necessary. Resource and scenario planning tools enable the company to distribute highly skilled employees very effectively. Especially when projects have to be prioritized or when resource bottlenecks constrain the workflow, this toolset facilitates reallocation decisions.

The next stage deals with the real product design and engineering process. The technical properties of the product are defined during this phase. It can consist of mechanical and electric components as well as software, depending, of course, on the product being developed. All engineering data is collected and enriched. During this process, virtual prototyping and simulations are used to determine whether the engineering matches the features and the requirements of the product. A clear link between engineering data and the specific requirements and features is always given.

The engineering data determine which parts and materials will be necessary to design the product. The engineering bill of materials (BOM) is thus automatically generated and flows into the company's ERP system. Management thus can live-monitor all product development processes. As decisions on the location of the plant have yet to be made, the engineering BOM is still location independent. This corresponds to the Siemens PLM Software Suite®'s general principle of "design anywhere—build anywhere". The next step enriches the existing data on design and the engineering BOM by plant specific data. These data consist, for example, of lists of material needed for manufacturing the products or the quantities of consumables and auxiliary materials for the production machines. It also contains further information on the specific resources for the particular manufacturing plant. E.g. the resource allocation to any task and sub-task can be determined individually.

Objects concurrently generated are the material routing information and plant data. The plant data can be determined by plant simulation, where the throughput parameters and machine line layout are generated. Also, the logistics are determined and simulated here. The generated data is merged with the manufacturing BOM to generate the material routing information. This information determines which materials are processed when and where. Specifications of the material and machines used are compiled here. The material routing information can be directly passed on to the manufacturing operations. The material routing plan as well as the manufacturing BOM enter into the resource library. Information on which material is needed at which points in time and at which part of the plant are merged and transferred to the ERP system. This is necessary for comprehensive logistics and budgetary planning. The data generated is seamlessly integrated into manufacturing automation and manufacturing execution. The step towards automatically transferring information directly into manufacturing execution systems (MES) is a further step towards establishing the concept of the "digital enterprise". Material routing information and 3D engineering data are automatically fed into the CNC programs used in machine tools. The post-sales documentation is automatically generated by the PLM suite during the product development and manufacturing process. This includes product manuals and maintenance instructions. As it is directly integrated into the flow of information, changes in product design or functionality will always trigger changes in the product documentation.

4.2 Implementation of the Siemens PLM Software Suite® at a Siemens Business Unit

In this section we highlight efforts as well as benefits in the period of implementing the Siemens PLM Software Suite®. To illustrate the effects, a case study is performed at a business unit within the Siemens Division Digital Factory. This business unit is on the third hierarchical level within the Siemens organizational structure.

The implementation of a PLM software environment is a time consuming process. For the Siemens business unit, the first steps of implementation started in 2009. First of all, the NX suite of the PLM software which comprises the CAD tools was implemented. These are the core parts of any product development environment. One year later, document management was implemented. The document management is most often the first step when creating a common data backbone. In 2011 all supplementary CAD development tools were integrated into the PLM environment and in 2012 the legacy PDM system "CADIM" was removed completely. "CADIM" was the previously used PDM-tool within the business unit. The production and manufacturing processes also used "CADIM" until 2012. A 3D process chain was established reaching from R&D to the suppliers and the manufacturing departments. All involved parties got access to necessary information while adhering to the principles of information security. The intricate authorization management system, which ensures high levels of data security, is also part of the Siemens PLM Software Suite®. Within the following 2–3 years, the requirements management and engineering tools including testing tools were also planned to be implemented. Consequently, this case study cannot provide the retrospective analysis of an entire PLM implementation. Rather, it can provide a review of the first implementation years and investigate the efforts, benefits and findings during this period.

Implementing the Siemens PLM Software Suite® means that the organization has to undergo an extensive, costly and often time consuming change process. We also observed this within the Siemens business unit. It is commonly estimated that the costs related to that change process amount to 50–150% of the cost incurred by the TCO of the actual software. Evidence of the case study shows that the time constraint related to the implementation is not caused by the installation of the new IT systems, but by the organizational change the company can handle without risking malfunctions. That is why we regard the range of 50–100% cost markup for the change process reasonable.

Concurrent to the implementation of the software, the processes within the company have to be adjusted to the standardized PLM environment. An appropriate and proven procedure to achieve this is value stream mapping and value stream design. Processes according to the standards of the software are a prerequisite for the implementation. Without this adjustment, some of the benefits such a software package offers will not be realized. A value stream visualizes all steps necessary to process raw material or components into a finished product which can be sold to a customer. It depicts all flows of information and material that finally lead to the sale of a finished product. McManus (2005) suggests three steps in the process of creating value stream maps: (1) Process steps have to be arranged, (2) performance data of the different tasks have to be collected and (3) how value is created must be analyzed (see McManus 2005, p. 37). Value creating and non-value creating steps are outlined. Redundant processes and wastes of time can instantly be identified and consequently eliminated. Depending on the type of product, the development

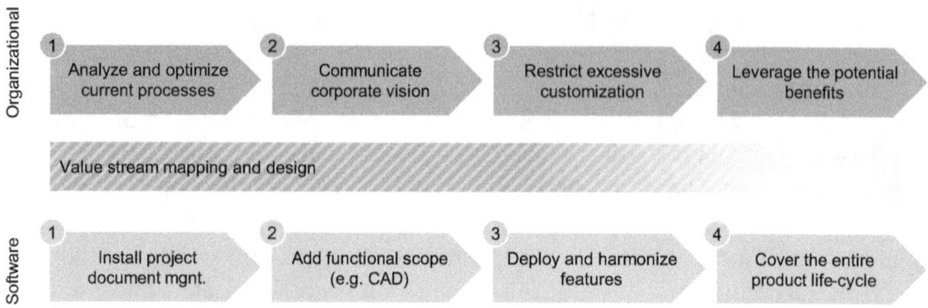

Fig. 7 Organizational and software based implementation of a PLM environment (Source: Own illustration)

process makes up a considerable part of the entire value creation. Value stream mapping as well as the PLM software implementation are to reduce non-value creating time and avoid time wasted.

As depicted in Fig. 7, value stream mapping can be regarded as the communication medium between the human aspects and the IT aspects of the software implementation. It makes the processes of the organizational and cultural change compatible with the change in the IT environment. Figure 7 depicts all necessary steps for the correct establishment of a PLM environment. Generally, the implementation of the software has to be divided into organizational and IT process steps. The time needed for the implementation is determined by the organization and its ability to keep up with the change process.

Some organizations promote a culture of change where the implementation can easily be carried out and advanced at a rapid pace. Companies in the industry sector usually are more reluctant to change, however. Their production processes evolved as the result of constant optimization efforts. Consequently, management and employees are rather opposed to disruptive process changes. In these companies a fundamental change certainly needs more time.

The first phase concerning the organizational change is the optimization of end-to-end processes. Here unnecessary steps and delays will be evident first and process improvements will show instant benefits. After having analyzed the processes, the PLM toolkit is used to coordinate the different steps and to make them compatible with each other. The main challenge here is to establish a link between all major product development steps and the common data backbone. This is essential for the functioning of the fully integrated system environment and enables the analyses of processes and sub-processes that provide the company with a holistic view of its own development cycle.

The second phase of the general process mainly consists of communication and support. Here it is vital that the whole top management team gives its full support to the implementation process. A vision statement should be issued by the top management team and be communicated to all employees, especially to the key stakeholders of the implementation project. These stakeholders have to be identified and then convinced of the benefits of implementation.

The optimization of the overall end-to-end processes should be especially emphasized. Local optimizations of sub-processes sometimes have to be sacrificed to enable the end-to-end process optimization. Benefits will thus be evident when looking at the overall result, even though local sub-processes might require more efforts.

When all key employees are informed about the project and convinced of its benefits, the information should cascade down the whole organization. As the organization not only has to bear but to support the change, it is of vital importance that no element of the system corrupts the change process. After the communication of the vision has been initiated, the workforce should be encouraged to support the change. After introducing the change, project tasks and functions have to be defined. It is important that everybody involved in the process knows exactly which role they play and which tasks have to be completed to ensure the success of the implementation project. During the implementation it is important to communicate the advantages to all stakeholders, for example that collaboration would be easier.

The third phase is characterized by software adaptation measures. As stated above, every implementation process is unique and different levels of customization are necessary. However, too much customization will reduce the benefits of the software. Accordingly, the standardized processes of the software should not be amended too much. Early in the process, the software requirements should be defined and subsequently modified. Milestones shall be defined with necessary software freezes. This limits excessive software development and lets everybody in the process concentrate on the most important tasks required to support the project.

During the last phase of the general process, after the software has been successfully implemented, the benefits can be leveraged. Strong partnerships with suppliers and customers lead to integrating their ideas and systems into the PLM environment. Early adopters of the system and its functionalities should be especially appreciated and supported. They can give valuable feedback not only on performance and functionality but also on software deficiencies. Once the software works well, they will be the promoters within the organization and will accelerate the pace of adoption.

The IT implementation should be carried out hand-in-hand with the organizational change process, in a kind of symbiosis, each one depending completely on the other. This means that in order to leverage the full potential of the change process, the organizational change has to be supported by computer aided systems that enable standardized processes and a common data backbone. An appropriate means to facilitate the implementation is the training of special key users. These key users combine expert knowledge on the software and simultaneously spur the mindset change of their colleagues.

In general, the first step of the software implementation should be limited in functionality like the installation of the project document management. Furthermore, it should only have a limited number of users to enable an implementation led by key users. This will greatly facilitate the implementation process. Subsequently, the range of the limited functionality package is enlarged to other departments involved in the product development process.

As soon as the documentation backbone works properly, functionality can be added to the product development process concerning all CAD tools of the mechanical and the electronic parts. These are commonly regarded as the core parts of the product development cycle; thus, it is important to integrate them early. Subsequent to the computer aided design integration, the remaining software features are implemented. This includes the early stages of the idea management and the requirements engineering.

In the fourth phase, the whole lifecycle of the product is covered by the software suite. The manufacturing planning as well as manufacturing execution applications are integrated and the transfer of CAD data to the CNC systems is enabled.

4.3 Costs and Benefits Related to the Implementation of Siemens Product Lifecycle Management

The most direct cost related to the implementation of the PLM environment is the licensing of the software. For the Siemens PLM Software Suite® there are over 50 types of licenses which are refunded by four main user-categories, the "Info-user", the "Project-user", the "Designer-user" and the "Author-user". These indicate the rights, authorizations and features the specific user has. To facilitate the cost analysis concerning the licensing of the PLM software, the license types were broken down to four main user groups. These are listed in Table 4 below with the cost per individual license given on an indexed basis. Each cost includes the single user licensing fee, the maintenance and support as well as the continuous software updates.

The "Info-user" license displays all information within the Teamcenter® data backbone. For example, employees at a production machine can review the design and feature documentation when they are unsure about the specifics in 3D production data. The "Info-user" cannot create new objects or amend documents within projects. He can, however, be integrated into existing workflows. This is the most common user group, thus leveraging the ubiquitous information potential of the common data backbone. The "Project-user" has the rights to create new data objects

Table 4 Yearly cost of ownership of Siemens PLM user

Summarized license type	Attributed cost per year indexed by "author-user" (%)	Distribution in business unit (%)	Absolute amount
"Info-user"	5	61	1404
"Project-user"	32	5	107
"Designer-user"	45	21	502
"Author-user"	100	13	290

Source: Siemens company data

and amend project documentation. These are usually project or product managers who create and monitor the progress made in development processes. The "Design-user" is additionally able to authorize designs and engineering of products or components. They can use the full functionality of the Siemens PLM Software Suite® with the exception of the CAD design features. These are only provided within the "Author-user" license. It comprises all CAD, CAE and CAM functionalities the Siemens PLM Software Suite® offers and is the core toolset within the PLM environment.

To track the indirect cost of the PLM installation, surveys were made to estimate the losses in productivity. The amounts which account for cultural change, suboptimal processes and process quality were analyzed.

The implementation of a comprehensive PLM system will lead to the abandoning of optimized sub-processes, enhanced process quality and cultural change.

The abandoning of optimized sub-processes has to be made for the benefit of a holistically optimized process over the product's life cycle. This will not only lead to a loss in productivity on the sub-process level, but also lead to initial opposition by the workforce, commonly regarded as cultural change. An example would be work with hot-key combinations when a new bill of materials is created. With the Siemens PLM Software Suite® employees have to click through a graphical user interface (GUI), thus more time is needed. However, these costs are continuously reduced by software improvements and updates as well as trainings and the natural learning curve of users. Increases in process quality, based on enlarged workflows within the new software environment, also cause indirect costs. The increases in process quality can be regarded as intentional additional expenses, necessary to develop superior products. To guarantee process security and transparency, check-in and check-out procedures are added if amendments are made to product design or documentation. These additional steps lower the productivity of sub-processes, but at the same time, the overall process efficiency as well as the quality of the whole process can be enhanced.

Figure 8 shows blocks indicating the productivity losses due to cultural change, the enhancement of process quality and the abandonment of optimized sub-processes. The productivity loss due to suboptimal processes and increased process quality was estimated to amount to 3.5%, including 0.5% accounting for cultural change.

These costs caused by the productivity loss will be gradually reduced as the software is continually adapted to comply with the processes within the business unit. This removal of sub-optimal processes amounted to 4.85% after the first year of implementation and thus more than offset the initial productivity losses. However, when PLM is installed and systems are integrated, there will be a temporary performance decrease. This is caused by the technical complexity of the holistic system in comparison to stand-alone programs without integration. These negative effects on the cost were offset by functional innovation and process innovation, which in sum amounted to 8.5% productivity increase after the second year of implementation. Additionally, productivity benefits are to be caused by the

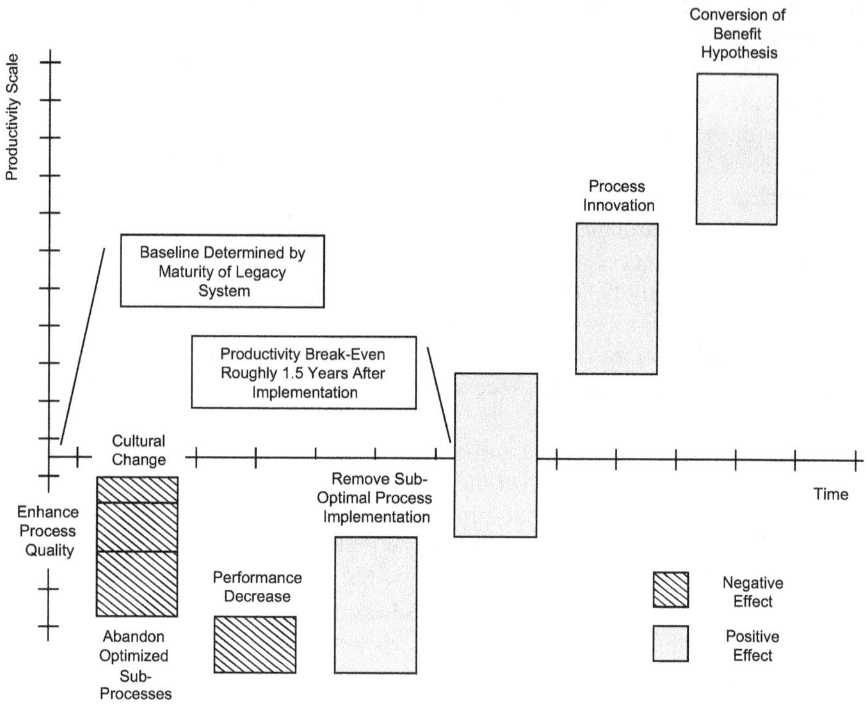

Fig. 8 Balance of productivity effects related to the implementation of the Siemens PLM Software Suite® for the Siemens business unit (Source: Siemens company data)

conversion of the benefit hypotheses which will be explained in the subsequent section. Within the Siemens business unit of this case study, the negative productivity effects caused by the PLM implementation could be balanced out one and a half years after the project start. However, there is a huge initial cost related to the implementation of a PLM environment. Whereas the productivity-based break-even can be realized after one and a half years, the accounting-based break-even will not be reached before a time horizon of 3–5 years. The decision to implement a holistic PLM environment thus has strategic dimensions.

To verify the benefits generated by the implementation of the Siemens PLM software suite, a monitoring system was implemented in the business unit. Tool packages and features of the software lead to savings in work time. The quantification and valuation of work time is the only possible method to measure the benefits of the PLM software implementation in a granular and timely fashion. The effects of a shorter time-to-market cannot be isolated. There are always positive or negative accompanying effects, which make a differentiated evaluation impossible. The product-to-market fit cannot easily be tracked either. A comparison would have to be made between two identical development projects. This is nearly impossible to fulfill, as product developments are made continuously. Parallel developments of identical products within traditional and new environments would be a waste of resources. That is why we are left with only the opportunity to analyze whether savings in work time and consequently costs could be realized.

The monitoring system to track these cost savings is led by a process manager for the Siemens PLM Software Suite® who supervises the implementation of software features within the business unit. After the implementation, the effects on workflow efficiency can be traced directly. An immediate comparison with the state before the implementation is possible in that case and thus the direct relation to the Siemens PLM Software Suite® is obvious. This monitoring system reveals negative as well as positive effects of the software. When the workflows are analyzed, the process manager makes a calculation of improvement where he lists the time savings in processes and describes how they were achieved. The improved processes and their frequency are presented to the head of the relevant department. Subsequently, the head of department reviews the calculation and confirms the validation with a signature. If the calculation is not signed, the process manager has to adjust previously made calculations.

4.4 Value Realized by Product Lifecycle Management at the Siemens Business Unit

Generally, it has to be said that numerous benefits that the Siemens PLM Software Suite® offers are not easily traceable. As explained in the previous section, there is a considerable cost associated with the implementation of a PLM environment. Thus every company has to analyze its specific situation to decide if an implementation makes sense. The benefits, listed in Fig. 9, are case specific for the business unit analyzed within this case study. Customers of the Siemens PLM Software Suite® have reported a large variety of outstanding benefits which helped them to secure and improve their market position. A sports airplane manufacturer could shorten the development cycle of a new plane from 6 to 3.5 years. In another case, a consumer goods conglomerate could shorten time-to-market by 25%, mainly by faster production line planning. A consumer electronics manufacturer could decrease development time by 30% mainly by using simulation tools instead of physical prototypes resulting in fewer defects in first series products. These examples show that different kinds as well as different values of benefits can be realized, depending on the specific type and situation of the customer.

Figure 9 shows a list of benefit hypotheses of a PLM environment within the case study's business unit. All hypotheses named can have a positive influence on the time-to-market, productivity, data quality, product quality, traceability, transparency, and cost.

The most influential levers to generate benefits within the analyzed business unit are global and engineering collaboration, the NX® tools for drafting, modeling and drawing, the integrated workflow system, the common data backbone and the CAD-CAM integration. These levers shall be explained in more detail.

1. Global Collaboration / Engineering Collaboration
2. NX® Drafting, Modeling, and Drawing Feature
3. Integrated Workflow System with Direct Data Link
4. Common Data Backbone
5. Less Programming Costs with CAD-CAM Integration
6. Simultaneous Engineering
7. Design Validation
8. Supplier Integration
9. Mechatronics Process Management
10. Integrated Portfolio Management
11. Project Document Management
12. Integrated Requirements Engineering
13. MS Office® as User Interface to Teamcenter®
14. Using Teamcenter® as Unique User Interface
15. Integration of ECAD Authoring System
16. Less Duplicates by Using Platform Management
17. Tecnomatrix® -TC® -Integration
18. T4S® - Interface (Teamcenter® - SAP®)
19. Less Duplicates by Using Geolus®
20. Manufacturing Management in One Tool
21. Higher Quality in all Product Developments
22. End-to-End Tracability
23. Documentation for Regulatory Compliance and Verification
24. Reporting and Analytics Function
25. Free Text Search Through all Documentation
26. Cross Reference for Norms and Certifications
27. Less License Costs for Different Applications
28. Less Administration Costs for Different Applications

Speed up Time-to-Market

Increase User's Productivity

Quality, Traceability, Transparency

Cost Savings

Fig. 9 Summary of benefit hypotheses of the Siemens PLM Software Suite®. All Benefits are only shown once over all categories, regardless of multiple effects in different categories (Source: Siemens company data)

Global collaboration means that all data and information on the products is available within the data backbone at any place and any time enabling collaboration across facilities in different locations in the world. For data protection, however, there is an intricate authorization feature inherent to the software. Dependent on their authorization status, users are provided with the rights to make changes to the necessary data sets or they are only allowed to display relevant information. All in all, it represents an optimized system that not only enables the implementing company to make use of the potential of outsourcing in developing countries, but also accelerates developments which are organized adhering to the "follow the sun" principle. It thus saves considerable costs and speeds up the product's time-to-market.

Engineering collaboration goes hand in hand with the global collaboration. It enables team members globally to view, mark-up and approve multi-CAD designs in virtual design reviews without requiring them to have a CAD license or CAD authoring experience. Thus, decision making processes are speeded up considerably. Reviews can easily be made with all relevant data needed which is important for CAE as the CAD models have to be enriched with data on engineering and manufacturability. In traditional product development a lot of communication effort between the departments caused many delays. With the Siemens PLM Software Suite®, every step, design and document is transparent. Consequently, there is less communication which increases overall productivity.

The NX® modules of the Siemens PLM Software Suite® are superior CAD tools. It is possible to use the sketcher data directly in the CAD drawings. The dimensions of the drawing are automatically generated. Compared to the previously employed software solution, the same results are obtained with an 80% reduction in workflow steps. HD3D[2] Visual high resolution product design displays facilitate decision making processes. The interface is optimized for creative tasks resulting in a 20% productivity increase of the designers' work.

The common data backbone is the main characteristic of a PLM environment. In traditional product development different software tools are available for the development process. The data and information generated is transferred to the next system when reaching the next development step. This procedure is time consuming as well as prone to errors. The common data backbone means that every department uses the same data—it can consequently be transferred to the next system at any time or when needed. All previous steps are transparent and traceable as every kind of information is displayed in one product tree. This feature also simplifies the handling of product variations. With the Siemens PLM Software Suite®, multiple product variations are managed by keeping track of which components are common to the platform and which are unique to the specific variation. This feature saves costs and speeds up time-to-market by reducing lag times in development and changes of products.

[2]High definition and three dimensional.

With the Siemens PLM Software Suite®, the CAD code is automatically translated to machine control programs. This means that there is a direct link from CAD design to CAM simulation and from there directly to the production machine without any programming. Time and costs are saved as no CNC programming is necessary and thus the time-to-market is accelerated. Formerly, CAD designs of products had to be translated into CNC programs for the manufacture of new products, which consumed time and resources.

4.5 Summary and Discussion

"All truth passes through three stages. At first, it is ridiculed. Secondly, it is violently opposed. Thirdly, it is accepted as being self-evident."[3]

When a new software suite is implemented which revolutionizes incumbent work processes, the mindset of people and the organizational culture have to pass through the three stages of truth identified by Arthur Schopenhauer. When the decision to implement a PLM software is made, it will be ridiculed. At first, the negative aspects of the change are evident, that is, the direct and indirect costs related to installing a PLM environment. It becomes clear that considerable effort is necessary to migrate data and to train employees in the new processes in the new work environment. Many employees and involved departments will ridicule the decision and predict the early death of the implementation project.

Dependent on the previously established processes, the second stage of social motion within the company will be violent opposition. When the engineers see that parts of their freedom in developing products is lost, they will pledge to restore the old system. Within the PLM environment they are obliged to stick to standardized development and approval processes which are necessary but uncreative work for them. Also employees who were for instance trained on optimized hot-key product data management systems will refuse the implementation, as they will see a loss in their productivity. Instead of setting up a bill of materials quickly by a combination of hot-keys, they now have to click through a GUI. This change and the inherent loss of productivity is comparable to the transition from DOS to Windows. Work processes within themselves will take longer when there is a GUI, but the system as a whole will have new capabilities that the previous one could not offer. Nevertheless, many employees will primarily see the loss in productivity in their personal scope of work and heavily oppose the implementation of the PLM environment. Many training sessions, management support and efforts at persuasion are necessary during this implementation stage. When these efforts are made and the employees start to realize the benefits of such a software suite, they begin to see the decision to implement a PLM environment as self-evident. The shorter development cycles, the easier global collaboration, the commonly transparent data, documentation access, and the advanced functionalities are perceived to outweigh

[3]Citation attributed to Arthur Schopenhauer, philosopher 1788–1860.

the productivity losses in singular work processes. At this stage, the implementation transitions into a mode of continuous improvement. As soon as the whole concept is accepted within the workforce, there will be positive self-enhancing dynamics. Employees will have ideas for further improvements and applications themselves.

In this chapter we have primarily characterized product lifecycle management and shown how it influences the value chain of a company. Then we outlined which benefits can be realized by the implementation of product lifecycle management. The main levers to increase company value are to achieve a shorter time-to-market, a better product-to-market fit and lower development costs. Shorter times-to-market lead to improved competitive positions: higher prices can be demanded and sales are facilitated when there is no competitor product in the market. A better product-to-market fit means that the features and the quality of the product meets the expectations of the customers. This quality can be achieved by early customer integration, requirements management tools and supplier integration. Lower development costs can be obtained due to automated processes, less communication effort between departments, and advanced outsourcing capabilities.

Subsequently, we showed that there are costs related to the implementation of a PLM environment. These costs consist of direct hardware and licensing costs as well as indirect costs stemming from efficiency losses. In the case study within the Siemens business unit we tried to compile the benefits and costs related to the PLM implementation. We showed that direct cost can easily be traced as there are bills and accounts for these. The indirect cost of efficiency losses and the organizational change is hard to track. Estimations are that these sum to 50–100% of the incurred direct cost. The case study demonstrates, that there is a huge initial investment necessary to implement a holistic PLM environment. After roughly 2 years, the break-even was reached with not all of the benefit hypotheses realized thus far. We have listed all benefit hypotheses related to the implementation of the software.

Summing this chapter up, our case study illustrates the benefits and costs related to the implementation of a PLM environment. Significant efforts are necessary, but such an implementation is probably the only way an industrial company can maintain its competitive position. Even with the application of benefit measurement approaches, the positive effects of a PLM environment are not completely traceable. The decision to implement a comprehensive PLM environment is of entrepreneurial character and is aimed at maintaining or enhancing a company's competitiveness. We expect that the utilization of software based tools along the whole product lifecycle will become industry standard. The necessity of the PLM environment will become self-evident.

References

Barczak, G., Abbie, G., & Kenneth, B. K. (2009). Perspective: Trends and drivers of success in NPD practices: Results of the 2003 PDMA best practices study. *Journal of Product Innovation Management, 26*(1), 3–23.

Chandrasegaran, S. K., Ramani, K., Sriram, R. D., Horváth, I., Bernard, A., Ramy, F. H., & Wei, G. (1991). The evolution, challenges, and future of knowledge representation in product design systems. *Computer-Aided Design, 45*(2), 204–228 (2013).

David, J. S., Schuff, D., & St Louis, R. (2002). Managing your total IT cost of ownership. *Communications of the ACM, 45*(1), 101–106.

Durmuşoğlu, S., & Barczak, G. (2011). The use of information technology tools in new product development phases: Analysis of effects on new product innovativeness, quality, and market performance. *Industrial Marketing Management, 40*(2), 321–330.

Ehrlenspiel, K., & Meerkamm, H. (2013). *Integrierte Produktentwicklung: Denkabläufe, Methodeneinsatz, Zusammenarbeit* (Vol. 5). München: Hanser.

Eigner, M., & Stelzer, R. (2009). *Product lifecycle management: Ein Leitfaden für product development und life cycle management*. Dordrecht: Springer.

Ellram, L. M., & Sue, P. S. (1998). Total cost of ownership: A key concept in strategic cost management decisions. *Journal of Business Logistics, 19*, 55–84.

Elsener, M. (2005). *Kostenmanagement in der IT: Leistungssteigerung und Kostenoptimierung; Kostentreiber erkennen und eliminieren; total cost of ownership, outsourcing, service level agreements, benchmarking; Methoden zur Kostensenkung und Effizienzsteigerung*. Frechen: mitp-Verlag.

Ferrin, B. G., & Plank, R. E. (2002). Total cost of ownership models: An exploratory study. *Journal of Supply Chain Management, 38*(2), 18–29.

Geißdörfer, K. (2009). *Total cost of ownership (TCO) und life cycle costing (LCC): Einsatz und Modelle: ein Vergleich zwischen Deutschland und USA* (Vol. 7). Münster: LIT Verlag.

Li, J., Tao, F., & Thao, L. (2015). Big data in product lifecycle management. *The International Journal of Advanced Manufacturing Technology, 81*(1–4), 667–684.

Martens, B., Walterbusch, M., & Teuteberg, F. (2012). *Costing of cloud computing services: A total cost of ownership approach*. System Science (HICSS), 2012 45th Hawaii International Conference on IEEE.

McManus, H. L. (2005). *Product development value stream mapping (PDVSM) manual release 1.0*.

Meckbach, G. (1998). Study reveals organizations blase about TCO. *Computing Canada, 24*(43), 15–16.

Myers, B. A. (1998). A brief history of human-computer interaction technology. *ACM Interactions, 5*(2), 44–54.

Nambisan, S. (2003). Information systems as a reference discipline for new product development. *MIS Quarterly, 27*, 1–18.

Potter, K., Smith, M., Guevara, J. K., Hall, L., & Stegman, E. (2011). *IT metrics: IT spending and staffing report 2011*. Stamford: Gartner Inc.

Ross, D. T., & Rodriguez, J. G. (1963). *Theoretical foundations for the computer-aided design system*. Proceedings of the May 21–23, 1963, Spring Joint Computer Conference, ACM.

Srinivasan, V. (2011). An integration framework for product lifecycle management. *Computer-Aided Design, 43*(5), 464–478.

Stark, J. (2015). *Product lifecycle management* (pp. 1–29). Cham: Springer International Publishing.

Vila, C., Abdellán-Nebot, J. V., Albiñana, J. C., & Hernández, G. (2015). An approach to sustainable product lifecycle management (Green PLM). *Procedia Engineering, 123*, 585–592.

Wiesner, S., Freitag, M., Westphal, I., & Thoben, K. D. (2015). Interactions between service and product lifecycle management. *Procedia CIRP, 30*, 36–41 (2016).

Wild, M., & Herges, S. (1998). Total cost of ownership (TCO) – Ein Überblick. *Management, 1*, 34–38.

Zarnekow, R., Brenner, W., & Scheeg, J. (2004). Untersuchung der Lebenszykluskosten von IT-Anwendungen. *Wirtschaftsinformatik, 46*(3), 181–118.

Measuring Innovation Performance

Friedrich Walcher and Ulrich Wöhrl

1 A Claim for Measuring Innovation Performance

Innovation is one of the most important strategic and operational levers for gaining competitive advantage and generating organic growth in an unstable economic environment. Business growth and profitability depend on continuous acceleration in innovation. Along with this imperative for innovation comes the necessity to adequately measure and control innovation performance. You cannot manage what you do not measure. This old management principle underlines the critical role of measuring innovation performance in order to enable sound analysis and decision making.

However, the measurement of innovation performance is a highly controversial topic and complex management challenge. Recent studies conclude that there exist discrepancies in what firms are hoping for and actually gaining from their investment in innovation (Adams et al. 2006). A survey of over 2700 executives reveals that only around 50% of executives are satisfied with the return on innovation spending (see Andrew et al. 2009). There is a consensus that innovation should be measured rigorously. But less than half of companies do so. One of the most eminent reasons for this seems to be uncertainty over the choice of the right metrics. This is caused by management's fear that strictly controlling innovation performance hampers innovation activities and limits the necessary degrees of freedom of the R&D department. Also, measuring innovation is found to be of minor priority in nearly a third of surveyed companies. However, measuring innovation performance is critical when it comes to making decisions. Allocation of R&D resources,

F. Walcher (✉)
TUM School of Management, TU München, München, Germany
e-mail: friedrich.walcher@tum.de

U. Wöhrl
Technology and Innovation Management, Siemens AG, München, Germany

G. Friedl, H.J. Kayser (eds.), *Valuing Corporate Innovation*, Management for Professionals, DOI 10.1007/978-3-319-64864-4_4

portfolio decisions and identification of future growth opportunities are supported by sound knowledge on the organization's innovation power.

The purpose of this section is to investigate how innovation performance of companies can be controlled and assessed in order to achieve sustainable improvement of innovation power. Focus is put on a holistic perspective and a c-level rather than a product technological view. Section 2.1 starts with creating a basic understanding of innovation and its attributes. The subsequent section deals with the characteristics of the innovation process and assesses the value drivers of innovation performance. In Sect. 3, a framework for measuring innovation performance at the corporate level is presented. The major areas that determine innovation power are discussed and possible indicators and key measures for innovation power described. The theoretical assessment of controlling innovation performance is complemented by a case study on the assessment of innovation power at Siemens. This case study provides an in-depth insight into the innovation review process at Siemens and establishes a link between theory and practice.

2 Fundamentals of Innovation and Innovation Performance

2.1 Characteristics of Innovation

Definition of the Term Innovation

There is no consistently used definition of the term innovation. This is due to the diversity of the innovation phenomenon and the heterogeneity of existing research with regard to sample and methodology. The respective objective of the author is decisive for the definition of the term innovation. Schumpeter, who contributed greatly to the study of innovation, defines innovation as the carrying out of new combinations (see Schumpeter 1934). This means that innovation is characterized by a new means-ends-combination, i.e. the use of new technology to meet a demand or need (see Gleich 2011, p. 59). Based on this idea, Utterback developed his approach of technology push and demand pull, that sees technology and customer demand as important drivers of the innovation process (see Utterback 1971).

The common consensus of existing definitions is that innovation is the "ideation, development and introduction of a new product, service, or process through a certain business model into the marketplace, either by utilization or by commercialization" (see Gamal et al. 2011, p. 7). Thus, innovation goes beyond pure invention or research and development, but also comprises the implementation of a marketable product and exploitation of its market potential in order to strengthen the competitive advantage of a company. This definition represents the complex and multidimensional character of innovation that makes measuring innovation performance a non-trivial matter.

Attributes of Innovation

The complex nature of innovation becomes obvious when taking a closer look at innovation's attributes. These are essential for understanding the difficulties of measuring innovation performance and for developing appropriate measurement approaches and metrics, as will be done in Sect. 3. Stone et al. (2008) determine nine key attributes of innovation. These attributes give indications for appropriate assessment areas of innovation performance.

1. Innovation involves the combination of inputs in the creation of outputs.
2. Inputs to innovation can be tangible and intangible.
3. Knowledge is a key input to innovation.
4. The inputs to innovation are assets.
5. Innovation involves activity for the purpose of creating economic value.
6. The process of innovation is complex.
7. Innovation involves risk.
8. The outputs in innovation are unpredictable.
9. Knowledge is a key output of innovation.

Some of these characteristics of innovations will be discussed in more detail. Knowledge as a key input to innovation contains people as an important factor for innovation (see Arvanitis et al. 2015). Entrepreneurial spirit and technological creativity of employees are key prerequisites for innovation activities and a fruitful innovation culture (see Kuratko et al. 2015). Thus, innovation is not just a complex management task but is often driven by single persons, i.e. visionaries who fully understand their market environment and intuitively know what is technologically feasible.

The eminent risk of innovations results from their novelty which goes in hand with entering new, unknown fields of expertise (see Hauschildt 1992). This risk is twofold: technical and economic (see Stippel 1999). The technical risk lies in uncertainty of whether development of a specific idea is possible. This ranges from inadequacies in employees acquiring the right knowledge to technical problems in prototyping and realization of production. Reasons for economic risk are based on a possible failure of the product in the target market, insufficient research on market potential or unpredictable customer reactions. Furthermore, innovation processes are characterized by detours and wrong turns as well as an oftentimes long process duration. Thus, information necessary for today's decisions often refers to events in the remote future.

Innovation assessment looks for indicators to identify and minimize these different types of risk. The eminent complexity of innovation ensues from the interlinking of a variety of different subtasks and departments, which is necessary for new product development. The innovation process itself is not linear, which adds to complexity. Rather, it is characterized by non-linear dynamics where inputs and outputs of iterative process steps influence each other. Novelty, complexity, uncertainty and risk are thus closely linked. Figure 1 gives an overview on these links.

Fundamental to the concept of innovation is the company's intention to create something of economic value. Innovation therefore can be seen as an intangible, risky investment. Expected cash outflows and inflows are uncertain and spread over a longer period of time. However, investment in innovation opens the opportunity of above-average returns (see Littkemann 2005).

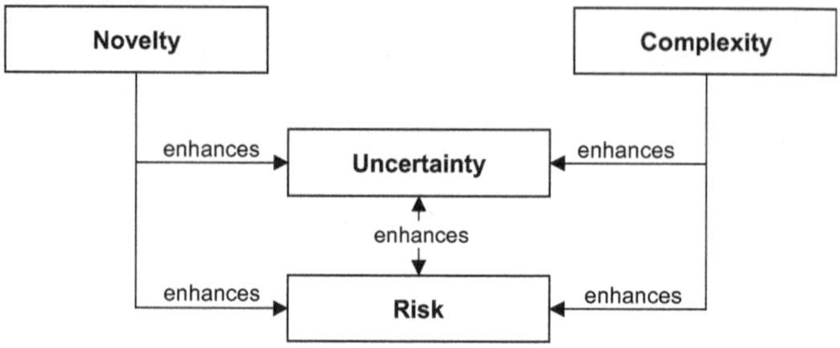

Fig. 1 Connections of innovation attributes (Source: Gleich 2011)

When it comes to measuring innovation performance, an important characteristic of innovations is the complexity in the accounting of innovation activities and innovation success. Compared to other functional areas of a company, it is more difficult to quantitatively ascertain inputs and outputs of innovation. This is due to several factors. Expenditures are not directly linked to income, which results in difficulties in profit-center building. Also, deriving a value contribution differentiated by functional areas is challenging since allocation of monetary benefits to specific actions is hardly possible. Even though in the development phase, innovation projects can mostly be seen as coherent systems, their economic effectiveness often results from a combination of different factors, for example, the combination of the respective innovation with services or other products of the company. Departments like marketing, sales and distribution also play their role in the success of an innovation. Thus, innovation is an interdisciplinary project that involves different segments of the company.

Time consuming development phases often result in long durations of the innovation process. This raises questions about the time interval of innovation performance measurement.

All these factors add to complexity in controlling and measuring innovation profitability and performance. It can be concluded that classical measuring methods and instruments which mainly focus on the level of R&D intensity are not suitable for this task.

2.2 Approaches for Assessing the Innovation Process

Not only innovation itself, but also the innovation process itself is of a complex nature. The uniqueness of each innovation is often referred to. Nevertheless, a broad standard process of innovation can be defined. A variety of different approaches exist in this field. We present two existing concepts that describe the innovation process in the following section: The innovation value chain of Hansen and Birkinshaw and the input-process-output-outcome approach of Brown and Svenson (see Brown and Svenson 1998; Hansen and Birkinshaw 2007). The concepts differ in their focus. The input-process-output-outcome model is especially suitable for describing the relationship of inputs and outputs and can therefore be used to analyze and assess the efficiency and effectivity of the innovation process. The innovation value chain approach focuses on the different value drivers and levers in the innovation process and is therefore a good basis for all qualitative indicators that should be part of an innovation performance measurement system. Both concepts are incorporated in the measurement framework that is described in Sect. 3.

2.2.1 The Innovation Value Chain

Hansen and Birkinshaw (2007), in their Harvard Business Review article, suggest seeing the innovation process as a value chain. The innovation process is broken down into three phases: idea generation, conversion, and diffusion of developed concepts. The concept of the innovation value chain represents a sequential, end-to-end view of the innovation process from accessing and creating knowledge, developing and building innovation, and commercializing those innovations. Across these three phases, managers must perform six critical tasks: internal sourcing, cross-unit sourcing, external sourcing, selection, development, and company-wide spreading of the idea. Each of these tasks represents a link in the value chain. The overall innovation performance of a company can only be as strong as the weakest link in this value chain. Figure 2 illustrates the three phases of the innovation value chain together with the six critical management tasks.

Idea Generation Innovation always starts with good ideas. These can come from within the own functional group, for example, the R&D or the sales department. Collaboration and brainstorming across business units combines insights from different parts of the same company in order to develop and create new ideas, products and businesses. The organizational structure plays an important role in benefiting from this lever. Geographical dispersion and a decentralized organizational structure often hamper employees from collaborating across units (see Hansen and Birkinshaw 2007). Sourcing knowledge from outside the company and also from outside the industry is an important part of idea generation. These external knowledge providers can, inter alia, be customers, end users, universities, competitors, investors, scientist and suppliers. Poor tapping into these external knowledge sources often results in lower innovation productivity and missed opportunities.

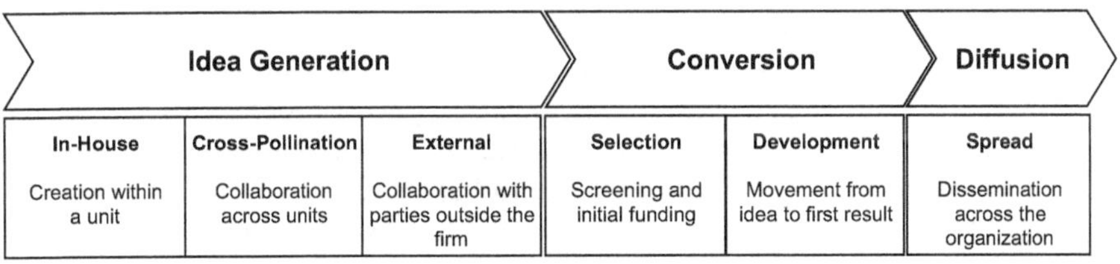

Fig. 2 The innovation value chain. (Source: Hansen and Birkinshaw 2007)

Idea Conversion Idea conversion involves screening and funding the ideas in order to transform them into codified innovations, i.e. new business models, products, processes or organization forms. Screening, i.e. prioritization, evaluation and focus, and funding mechanisms play an important role in this process step. If funding criteria are too strict and budgets are tight, this can shut down promising ideas. On the other hand, if screening and funding criteria are lax, the company can be overflowed with projects of low quality. The fit of the idea with the corporate strategy plays an important role in this context and must not be neglected in the screening and funding process. Section 3.2 will discuss this. Once selected, each idea must be transformed into a product, i.e. a revenue-generating service, product or process. Part of the conversion process, therefore, is to control and secure the viability of each idea in the funnel. This may include participating with external partners.

Looking at the innovation process as a value chain has one big advantage: it becomes obvious that strong performance in each part of the chain is indispensable for strong overall innovation performance. Thus, the innovation performance of a company is only as strong as the weakest link in the value chain. The innovation value chain seems to be an adequate representation of the innovation process with regards to measuring innovation performance on firm level. A company with many innovative ideas is not necessarily good at innovation performance. It may lack in conversion of its ideas and be a conversion-poor company. Similarly, an idea-poor company might be good at conversion and diffusion and therefore needs to rethink and improve its idea generation processes in order to foster overall innovation performance. The innovation value chain can help managers to identify the weaknesses in the innovation process and as a result can shift focus on implementing the right innovation tools and approaches to foster overall innovation capabilities. Thus, the value chain approach is especially suited for processing concrete innovation topics and the achievement of consistent innovation strategy.

Idea Diffusion In this process step, the concept has been sourced, funded and developed. Now the process of exploitation through which innovations are translated into sales gains or productivity gains starts. The new product must be spread across desirable channels, geographic locations, and customer groups. This link in the innovation value chain may include the use of intellectual property protection as well as different forms of customer involvement for building branding and reputation.

2.2.2 The Input-Process-Output-Outcome Model

The input-process-output-outcome model (IPOO) focuses on the critical evaluation of employed resources and achieved outcomes. Originally proposed by Brown and Svenson (1998), it displays an idealized structure of the innovation process (see Brown and Svenson 1998; Janssen and Möller 2011). Figure 3 displays this structure. The processing system, i.e. the R&D lab, transforms certain input factors in forms of people, ideas, equipment, funds and information by means of activities like researching, developing, and testing. The outputs of this transformation process

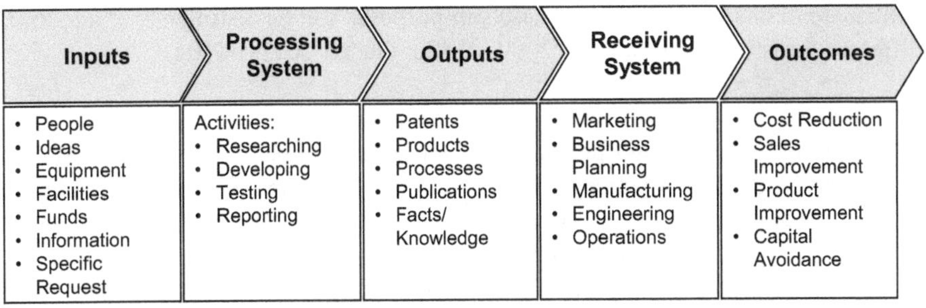

Fig. 3 The input-process-output-outcome model (Source: Janssen and Möller 2011, p. 99)

are new products, processes or business ideas. These outputs are then absorbed by the receiving system that includes all business departments which are involved in the commercialization of an innovation, for example, marketing, manufacturing, engineering and operations. Commercialization of the innovation outputs ultimately leads to quantifiable outcomes in the form of any combination of revenues, cost reductions, sales improvement, competitive advantages, and capital avoidance.

In delineation to existing input-process-output models like, amongst others, that proposed by Werner (2002), the IPOO model extends these models by the outcomes that represent the commercialization and exploitation of innovation ideas. Brown and Svenson stress that these outcomes should receive special attention. Hence, the IPOO model denotes a holistic illustration of the innovation system, covering all characteristic elements of innovations that were defined in Sect. 2.1.

Clearly, taking a strict process perspective is helpful in gaining understanding of the relations between inputs and outcomes, i.e. of the efficiency of the innovation process. Consequently, the IPOO-model represents an insightful framework for measuring the efficiency of the innovation process. Furthermore, this model shows that innovation performance is not only dependent on R&D performance alone but also on the contribution of other functional areas such as manufacturing, marketing or sales. Section 3.2 will describe this in more detail. However, a drawback of such a process perspective is that innovation is often more complex in practice and does not follow such a linear process scheme. Furthermore, contextual factors like culture and strategy that also play an important role in innovation performance are missed out (see Sect. 3.2).

2.3 Fundamentals of Innovation Performance Measurement

2.3.1 Reasons for Innovation Performance Measurement

Generally, the overarching objective of innovation performance measurement is ultimately to increase the efficiency and effectiveness of an organization's innovation activities. This is very vague, but having a closer look at some more specific sub-purposes will shed light on how measuring innovation performance can contribute to this ultimate goal. These sub-purposes can be distinguished into three different areas (see Chiesa et al. 2009, p. 496; Gleich 2011, p. 104):

1. *Information and Diagnosis*

   Measuring innovation performance is key for companies to understand their current innovation practices and innovation capabilities. In order to maximize innovation success, it is important to know about fields and business units where innovation capabilities fall short of competitors or management's expectations. Consistent measuring of innovation performance in on-going innovation performance audits provides management with past- and future-oriented control information. This is key for being able to discover, identify and pinpoint specific areas of weakness among an organization's innovation capabilities. Also, areas of strength to capitalize on can be identified. Developing a firm-level innovativeness index for the sector companies and business units allows monitoring of innovation success development over time. Benchmarking the organization's innovation performance with international top innovative companies helps to analyze the organization's own competitive position.

   Information on the innovation power of a company is also of interest for external stakeholders. Shareholders of public listed companies expect information on the innovation performance of a company to estimate future prospects of success. Innovative companies can profit from an innovation premium, reflected in a positive capital market perception, a trust bonus, higher market valuation and higher stock prices (see Trautmann and Enkel 2014). Having strategic innovation communication to foster positive perception and valuation of their innovativeness by the capital market is therefore of major importance for companies. Measuring innovation performance is one element required to succeed in this ambitious task.
2. *Coordination and Steering*

   Based on the identified areas of weakness, action can be taken to enhance innovation capabilities in the organization. Through measuring innovation performance, management's and staff's attention can be focused on the right projects, the most promising activities and the right actions from an innovation performance point of view. This helps controlling innovation activities and overcoming the barriers that stifle creativity and innovation. Knowing about innovation performance of the company allows the allocation of available resources in the most efficient way possible and influences decisions, e.g., divestment or investment decisions. Information on innovation power is thus a basis for strategic decision-making by top management. The case study will illustrate this in more detail (see Sect. 4.2).
3. *Motivation of employees*

   Measuring innovation performance and thus knowing about the status quo of innovation activities' performance within the organization is the basis for choosing the right goals and respective incentives to drive employee involvement and motivation. Besides, a firm-level innovativeness index spreads the awareness of the importance of innovation concepts and strengthens the innovation culture within the organization.

2.3.2 Levels of Innovation Performance Measurement

Innovation performance measurement of an organization can be classified into different levels (see Schentler et al. 2010, p. 305). These range from single innovation projects (single-project level) to the increasingly granular level on company-level:

1. Corporate level
2. Portfolio level
3. Single-project level

Depending on the level, the approaches for measuring innovation performance differ in focus and methodology. The measurement of performance on all three levels creates a detailed understanding of the innovation capabilities and activities for the company. Only an integral way of measuring innovation performance over all existing levels allows the capture of the different dimensions of innovation success (see Schentler et al. 2010, p. 306). As already elaborated above, we focus on measuring innovation performance on corporate level. However, it is important to know about the characteristics of the measurement approaches on each level. Therefore, the differences of measuring innovation performance on the different levels will briefly be highlighted below.

Corporate Level

The focus of measuring innovation at the firm level focuses on R&D effectiveness. "Doing the right things"—this is the ultimate goal of innovation management on corporate level. An innovation process is effective when its outcome is in accordance with the business objectives (see Bleicher 1990). Hence, of specific importance in this context is that the short-, mid- and long-term company strategies are in accordance with the innovation strategy. Consequently, controlling and measuring innovation performance at the corporate level goes beyond the mere addition of innovation performance on product and portfolio level. Instead, it measures achievement of contextual factors like strategic objectives and ensures that innovation outcomes are aligned with corporate culture. Companies increasingly see assessing their innovation capabilities and innovation management performance as part of measuring innovation performance at the corporate level (see Ochoa and Peña 2012; Schentler et al. 2010). This underlines that notion that measuring innovation performance at the corporate level is a prerequisite for putting innovation management into action and consequently is input for the optimization of the business portfolio (see Schentler et al. 2010).

Portfolio Level

The objective of controlling and measuring innovation performance at the portfolio level (equivalent to division or business unit level) is to maximize the financial value of an organization's innovation portfolio. Evaluation, prioritization, and selection of new R&D projects can only be carried out in a proper way when the organization has established a controlling and measurement process at the portfolio level. Otherwise, it is likely that a company allocates resources to ineffective innovation projects and the company engages in innovation projects that are not promising (see Balachandra and Raelin 1980, p. 24). Only when innovation performance at the portfolio level is measured constantly, can the right balance among the variety of different projects of the company be ensured and projects can be optimally selected according to the available capacity.

Single Project Level

Single projects can be evaluated at different stages as they evolve from the initial idea and concept to a final product, process or service. Each innovation project needs to be considered as a planning and controlling object. Performance measurement approaches at the single-project level focus on efficiency. Quantitative aspects can be timelines, design-to-cost, and meeting of target costs. Qualitative aspects are often the degree of innovativeness, stakeholder satisfaction, process quality, knowledge enhancement, and conformity to the organizational strategy (see Schentler et al. 2010, p. 313). Thus, the focus of measuring innovation performance at the single-project level is on operative excellence, execution and control of processes rather than on strategic aspects.

Performance measurement of the three described levels shows different focuses and follows different objectives. Organizations should control and measure their innovation performance on all three levels in order to gain a holistic view off their actual performance. Only then can effectivity and efficiency of the innovation activities be secured. Therefore, the performance measurement activities at the different levels must be linked. The specifications of the performance measurement activities should be synchronized and coordinated to cover all the relevant dimensions of innovation performance. Figure 4 provides an overview of the different focus areas of innovation performance measurement as well as their key objectives.

2.3.3 Characteristics of Innovation Performance Measurement Systems

The multidimensional character of innovation performance suggests that for measuring performance in the context of innovation, the simultaneous use of different performance indicators is necessary. These quantitative and qualitative measures should not be applied as loose and unconnected sets but rather be integrated into a performance measurement system. An innovation performance measurement system comprises all mechanisms that support the measuring process of innovation performance (see Möller and Schönefeld 2011, p. 368). It describes the operationalization of measurement and usually consists of several measurement procedures. Each procedure is designed to fulfil a certain purpose, e.g., to motivate a product development team, or to give managers insights to allow the undertaking of corrective actions. When setting up a performance measurement system, managers should never lose track of the purpose of every single measurement procedure. Otherwise, acceptance of the system and efficiency of the measurement process might be low. Each measurement procedure is characterized by a

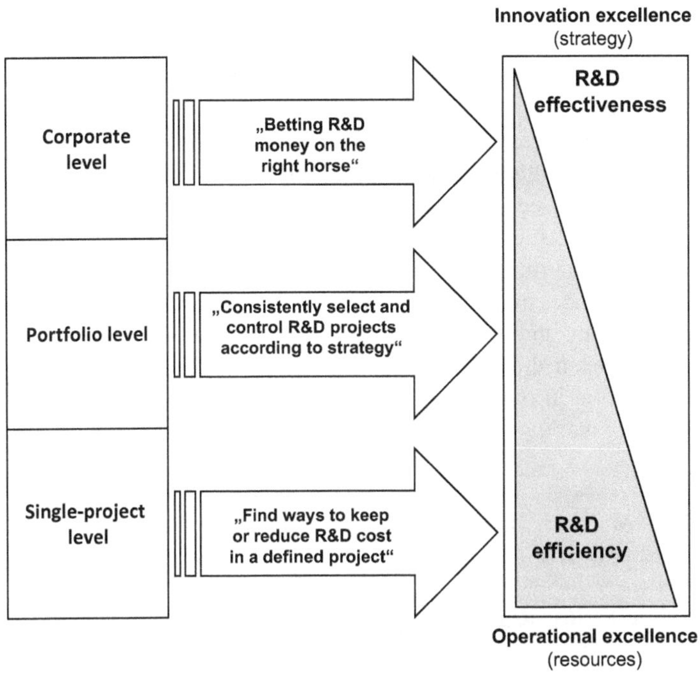

Fig. 4 Innovation performance measurement for different company levels (Source: Siemens AG, Technology and Innovation Strategy)

combination of metrics (performance aspects and indicators), the measurement methods, the frequency and timing of measurement and the reporting format. The innovation performance measurement system depicts the implementation and application of the different measurement procedures and combines a variety of performance indicators. A performance indicator is usually defined as a variable which indicates the efficiency of effectiveness of a system, part of a system or process when compared with a reference value (see Kerssens-van Drongelen and Cook 2000, p. 121).

Two basic distinctions can be made in order to classify possible measurement methods. The first is whether the method results in quantitative or qualitative metric values. For example, computational methods lead to quantitative values (e.g. number of patents granted). Assessment methods normally result in qualitative indications of the metric value (e.g. the quality of the innovation process has been 'satisfactory'). Qualitative values can be converted into numerical equivalents, for example, five-point scales. This conversion can be done by using techniques such as profiles, scaling models, check-lists and scoring models (see Werner and Souder 1997).

The second distinction that can be made for measurement methods is whether the method relies on subjective judgments or uses objective information. The degree of involvement of the evaluator in the subject of measurement and the number of evaluators is important in this context since the judgment of a group of experts is considered as more objective and credible than the highly subjective judgment made by one person directly involved in the subject of measurement. This is important to consider when setting up the measurement framework.

In the literature, multiple frameworks for measuring innovation performance at the firm level exist (see Adams et al. 2006). As outlined above, an important precondition for a holistic view of performance innovation is that the measurement system covers all the dimensions of innovation performance success. As Adams et al. (2006) conclude, most existing approaches fail in doing so. Derived from a synthesis of existing studies, they suggest a synthetic and integrative framework that comprises seven different measurement areas. This approach has received increasing attention in recent literature (see i.a. Möller and Schönefeld 2011). It will be presented and discussed hereafter.

3 Towards an Innovation Performance Measurement Framework

3.1 Basic Framework Design

In Sect. 2.2, two basic models of the innovation process were presented: the input-process-output-outcome model and the innovation value chain. It was elaborated that the IPOO model focuses on the process character and efficiency and takes into account the core aspects of innovation. The innovation value chain takes additional contextual factors into account and is especially suitable for the identification of weaknesses in the innovation process and the derivation of necessary actions. The innovation performance measurement framework according to Adams et al. (2006) consists of seven different measurement areas. Each category consists of a series of sub-dimensions (see Fig. 5):

1. Inputs
2. Knowledge management
3. Strategy
4. Organization and culture
5. Portfolio management
6. Project management
7. Commercialization

By focusing on the above listed seven areas for innovation performance measurement, the two basic aspects are taken up which makes the framework a holistic one: a focus on key activities and success factors of the innovation tasks and a consideration of quantitative as well as qualitative aspects. Strategy, organization and culture, and knowledge management cover the contextual factors that influence innovation performance and value creation. Inputs, project management, portfolio

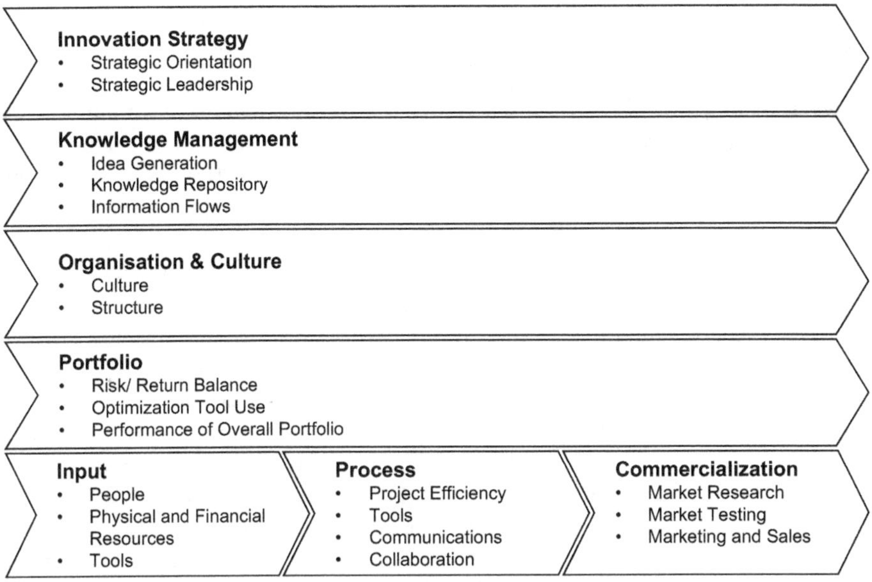

Fig. 5 Categories of an innovation measurement framework (Source: Möller and Schönefeld 2011, p. 369)

management, project management, and commercialization represent the sequential character and the different stages in the innovation process. Thus, the presented framework can help to overcome the prevailing problem that metrics within the field of innovation are mostly concerned with capturing financial outcomes (see The Boston Consulting Group 2009). This is because innovation activities aim to generate competitive advantages and create economic benefits with new products or services. However, as, e.g., the innovation value chain approach suggests, financial metrics on its own are not suitable for describing an organization's innovation performance. Figure 5 graphically visualizes the innovation performance measurement framework with its seven measurement categories.

3.2 Categories of the Measurement Framework

3.2.1 Input Management

Inputs denote the resources available for innovation activities. The amount, combination and quality of input factors have an effect on innovation performance (see Werner 2002). Inputs include the categories financial, human, physical and technological resources. Personnel expenses amount to the biggest share of input expense. According to a study of Grenzmann et al. (2004), 58% of R&D expenses in the German economy are personnel costs.

A frequently used global measure of input is the R&D intensity. It is typically expressed as a ratio between R&D expenditure or numbers of employees in R&D roles and an output measure, e.g., overall sales. The meaningfulness of R&D intensity needs to be considered with care. A number of studies have empirically examined the relationship between R&D intensity and innovation performance. The results are ambiguous. Whereas e.g. Stock et al. (2001) find an inverted u-relationship between R&D intensity and new product development performance, Bougrain and Haudeville (2002) cannot find any influence of R&D intensity on project's prospects. It is necessary to put R&D intensity into the company's specific context, i.e. company size, industry sector, business model and type of business (product, project, solution, service), market environment, process designs, etc. For small and medium-sized enterprises which often do not have formal R&D processes and activities, R&D intensity might not be an adequate measure for input efficiency. Companies in service industries tend to have low R&D intensity. High levels of R&D intensity also can be evidence of inefficient process design and not necessarily for good innovation practice (see Dodgson and Hinze 2000). Yet clearly, adequate funding is a critical input into the innovation process. Being aware of the equivocalness of R&D intensity is important when applying it, though. A possible solution to overcoming this obstacle is to combine this indicator with measurement outputs of the other six categories of the innovation performance measurement framework. Only then can the efficiency of the entire innovation process be assessed and a sound evaluation of the appropriate level of R&D intensity be undertaken.

Qualitative measures that aim to determine the adequacy of funding can further emphasize and improve the meaningfulness of pure financial measures for R&D intensity. Disaggregating the inputs of a company's innovation activities into different types, i.e. human, financial, technological and physical resource, seems to be the most promising possibility to gain a sound understanding of the quality and adequacy of the input level. A diversity of skills and experience of people participating in the innovation process can add significant value to innovation outcomes (see Damanpour 1996). Therefore, measuring a firm's innovation potential and performance should also consider employees' motivation, whether they show challenging behavior and contribute to the implementation of new ideas. This can be done via scales such as the innovative work behavior measure. This metric consists of different question spanning over the four areas: idea exploration, idea generation, idea championing, and idea implementation (see de Spiegelaere et al. 2012). The metric can be applied at the firm level as well as at the individual level.

The use of systems and tools, so called technological and physical input, is also an important input to the innovation process. It can be counted in quantified measures such as euro terms. Also, availability and use of tools.as well as the quality of input technology can be helpful when controlling innovation performance. In conclusion, there are several ways to assess the inputs to the innovation activities at the firm level. It is important that the developed set of measures is balanced and covers all sub-dimensions of the input categories (see Adams et al. 2006, p. 28).

3.2.2 Project Management

The innovation process itself, i.e. the transformation of inputs into marketable innovations, is in focus of project management. Having an efficient process that is able to deal with the heterogeneity of innovations is agreed to be critical to innovation performance (see, amongst others, Globe et al. 1973). Different concepts for visualizing the innovation process have been presented in Sect. 2.2. When it comes to measuring the performance of the innovation process itself, a company should evaluate three areas of project and process management: process efficiency, internal and external communication and collaboration (see Adams et al. 2006, p. 36).

Three dimension should be considered when evaluating process efficiency: cost, quality, and time. Comparisons between actual overall project and process costs are a good indicator of process efficiency at the firm level. Innovation speed is of specific importance in times of reduced product life cycles. Optimizing the time-to-market also helps to save costs and gain competitive advantages. This has been elaborated in detail in Sect. 3.2. Milestones and stage-gate concepts are specifically suitable for measuring time and quality components of innovation performance on project level, but can also be adapted to the firm-level.

There is a positive correlation between internal communication and innovation performance (see Damanpour 1991). Communication facilitates the creation and spreading of new ideas within the companies, increases the quality of ideas and helps to identify weaknesses of innovation activities in an early stage. Furthermore, it creates a climate that motivates and inspires employees. Hence, the existence of a sound communication within the functional areas involved in innovation activities is an important part of the innovation process and should be considered when controlling and measuring innovation performance at the firm level. Measurement of internal and external communication should focus on whether it takes place, with whom it takes place and the quality and level of the communication itself. Possible measures can be objective frequency counts as well as subjective evaluation. Especially the objective evaluation of communication practices plays an important role when controlling innovation performance at the firm level. If supervisors and management consider communication as rather unimportant or rate communication quality as poor, this can hamper future innovation performance of the company. Of special importance in this context is the communication across teams and functions.

The importance of collaboration has already be elaborated in the description of the innovation value chain in Sect. 2. Collaborating with customers and suppliers can improve the outcome of the innovation process. At the firm level, the quality and level of collaboration can e.g. be the percentage or number of projects that are developed in cooperation with third-parties and the intensity of the collaboration process (see Adams et al. 2006, p. 37).

3.2.3 Commercialization

By definition, commercialization is key element of any innovation (see Sect. 2.1). It aims at making the developed product or process a commercial success and includes issues like marketing capabilities, sales, distribution and joint ventures (see Adams et al. 2006, p. 37). Measuring commercialization in context of innovation means at the one hand measuring the financial success of an innovation in forms of outcome measures like revenues generated by specific technologies. On the other hand marketing capabilities and marketing performance necessary for commercialization need to be evaluated and assessed. Market success can be measured via numbers like sales, profitability or savings arising from innovations. Percentage of projects that are commercially successful or percentage of sales revenues from new products and services can also be adequate measures at the firm level. It can also be beneficial to gather information on the number of users or adopters of the new technology. A difficulty in measuring innovation outcome is the time-lag between start of innovation activities and first outcomes as well as the accounting of outcomes of special innovation activities.

For the success of any new product, marketing, sales and distribution capabilities are of utter importance. In practice, innovation activities and marketing activities are often regarded as strictly separated fields (see Möller and Schönefeld 2011, p. 370). For measuring innovation performance, companies need to overcome this separation and take up evaluation of marketing and sales performance in the innovation performance measurement system. Because only if marketing and sales capabilities are well-developed, is the transference of innovation into outcome ensured. Customers and market development should not be neglected when assessing commercialization performance. Therefore, the following key questions can be part of the measurement framework: are we positioned properly for changes in the beliefs, ideals and attitudes of our customers? How well do our products and sales processes match our customers' need? (see Morris 2008, p. 15).

Adams et al. (2006) conclude that firm-level measures for commercialization process are relatively poorly developed. In fact, the area of commercialization is the least developed of the dimensions involved in innovation performance measurement and there is need for further research regarding measurement viewpoints necessary in this area (see Möller and Schönefeld 2011, p. 371).

3.2.4 Knowledge Management

A company's ability to identify, acquire and utilize knowledge heavily influences a firm's success (see Zahra and George 2002). Knowledge is the central success factor for innovations and thus it is important for companies to measure their capabilities in this field. With regard to knowledge management, three sub-categories can be distinguished that are important for the innovation process: idea generation, knowledge repository, and information flows. Generating a sufficient number of high-quality ideas is the start for any innovation activity. At a firm-level, the number of high-quality ideas and the number of ideas generated that end up being selected and funded can represent possible measures for assessing idea generation processes. But companies should not only measure the quantity of ideas. First and foremost, the quality of any idea should be the focus of management attention. In this context, companies need to specifically define the term high-quality idea. Of central importance is also to evaluate the so called absorptive capacity of the firm, i.e. the ability to absorb and apply new knowledge. Strong absorptive capabilities make a firm more likely to acquire knowledge from outside and learn and benefit effectively from existing research, competitors and consumers. A frequently used indicator for explicit knowledge repository is the number of patents granted (total or in a given period).

Information flows within and into the organization are important for allowing the development of innovative concepts and for inspiring creativity. Three approaches for measuring information flows can be identified: first, measures of internal information gathering processes. These can be statistics on project reviews and statistical reports. Second, measures of the linkages that the company maintains with external sources and organizations. Typically, these are dichotomous measures, e.g., collaboration with universities through participation in research projects or attendance at trade shows and industry conferences. Third, measures for customer information contacts. Customer contact time and the adequacy and usefulness of information gathered via customers can be possible indicators for this category (see Kerssens-van Drongelen and Cook 1997).

3.2.5 Innovation Strategy

Innovation strategy describes the long-term innovation posture of an organization in terms of its new product and market development plans with regard to its competitive environment (see Sundbo 1997). It should be aligned to the overall business strategy of the unit. The indicators for this category that are embedded in the innovation performance measurement framework should focus on the essentials and avoid short-term orientation of the company. The innovation strategy is derived from the company strategy and should be aligned with the latter. Defining such an innovation strategy is a crucial element for the success of innovation activities (see Schentler et al. 2010, p. 306). It influences different kind of decisions. First, the selection of target markets, technologies, and products to invest in. Second, the pre-selection of ideas and projects in the organization's innovation field. Third, the allocation of resources that are assigned to each field of innovation. Fourth, it influences the composition of the innovation portfolio, the patent strategy, and the standards of an organization.

Based on the classic s-curve model, three general types of innovation strategy can be distinguished. The first mover, the trendsetter and the fast follower. The first mover is characterized by being the first to enter a specific market with a new technology, application or business model. Thus, the company is not only the ideas leader, but also sets the benchmarks in the specific field. The trendsetter enters the market at a later stage in the s-curve model. The trendsetter gains technology power in a market and sets standards in the respective field. The strength of a trendsetter company is to take up existing technologies and transform them into key-technologies. The fast follower avoids high entry costs that are caused by possible market and technology failure. He tries to enter an existing mass market not by setting new standards and being an innovative leader, but by price or cost leadership.

People/Skills/Culture; ...

Processes & Organization

Innovation Strategy	R&D Ressource Allocation	Technology strategy	Patent strategy	Standards
"First Mover" • Ideas leader • Benchmark breaker • Pilot customers	• Creative R&D • Freedom of action/ spin-offs • Use external idea pools	• New technologies (discontinuity, new rules of the game)	• Basic patents • Selectively strong patent portfolio • Potential for generating license revenues	• Work independently • Strategic partnerships
"Trendsetter" • Market power • Technology power • Several lead customers	• Effective R&D • R&D closely linked with business strategy • Strategic partnerships	• Pace-setting technologies (determine tomorrow's competitiveness)	• Own basic patents or access them • Key patents • X license agreements, freedom of action	• Dictate the (defacto) standardization process • Control further development at standard level
"Fast Follower" • Processes • Cost leader • Other core competencies	• Efficient R&D • Systematic make or buy	• Basic technologies (basic competence for today's business)	• Licensing • X license with compensation payment • Improvement patents	• Make quick use of available standards

Fig. 6 Example influences of the innovation strategy on different fields of the innovation process (Source: Siemens AG, Technology and Innovation Strategy)

As visualized in Fig. 6, the chosen innovation strategy has implications on R&D resource allocation, technology strategy, patent strategy and standards, which, inter alia, include process design, organization and culture. For example, R&D performance assessment of a company with a first mover strategy clearly needs to focus on creativity, whereas the trendsetter puts emphasis on effectiveness of R&D. In contrast, the fast follower places importance on efficient R&D and efficient processes.

The implications of innovation strategy on different measurement dimensions need to be considered when measuring innovation performance. Management of inputs, process, commercialization, knowledge management, organization and culture as well as of the overall innovation portfolio always takes place in due consideration of the innovation strategy. Therefore, the chosen measures in the innovation performance measurement framework must account for this.

3.2.6 Organization and Culture

The innovation culture is the combination of the innovation related attitudes, experiences and values of the employees in an organization (see Schentler et al. 2010, p. 306). Only if the overall culture in a company supports, coordinates, and drives innovations and creates a place in which they can take place, will the company be innovative (see O'Reilly and Rao 1997, p. 60). A strong innovation culture can compensate for less R&D intensity by creating a new style of corporate behavior that fosters creativity and new ideas and allows change, risk, and failure. Measures for the performance of the cultural dimension of innovation management are mainly qualitative ones (compare Schentler et al. 2010, p. 397 and Amabile et al. 1996, p. 1158). The encouragement of creativity within the organization, for example, risk attitude and rewarding practice for creativity, is one criteria in these terms. The propensity to take risk describes the willingness to confront risky opportunities, tolerate possible failures and learn from those. Freedom and autonomy are a prerequisite for innovative work and also important to measure at the firm-level. The measurement framework should be able to identify factors that impede creativity of employees and the organization. Furthermore, the adequacy of resource distribution and its effect on motivation is part of the innovation culture of a company and hence should find its place in a measurement framework. Motivation is an important factor, not only of the innovation process, but also of the incentive structure and reward system (see Adams et al. 2006, p. 34; Möller and Schönefeld 2011, p. 370).

The innovation organization and structure represents the innovation-related organizational aspects of a company. The innovation structure links structured activities with roles and responsibilities (see Schentler et al. 2010, p. 308). It needs to combine two conflicting dimensions. On the one hand, there needs to be enough freedom and flexibility in idea generation. On the other hand, innovation activities need to be structured and controlled in order to effectively and efficiently support success of innovations. Personnel and organizational flexibility can be measured with qualitative indicators such as 'corporate flexibility and responsiveness to change' (see Ekvall 1996). The adaptiveness of R&D personnel to technology changes is a measure proposed by Lee et al. (1996). Centralization, i.e., the concentration of decision-making at the top of the organization hierarchy and the degree of emphasis of authority on role performance have a negative impact of organization innovation (see Damanpour 1991). Both the degree of centralization and of emphasis on role performance, and decision processes of top authorities can therefore be helpful metrics in order to measure the performance of the structure-related dimension of innovation. Indicators such as 'freedom to make operating decisions or the 'degree of empowerment' are also regarded as suitable metrics for autonomy (see Abbey and Dickson 1983; de Leede et al. 1999).

3.2.7 Portfolio Management

The effectiveness in managing an organization's R&D portfolio is often a key determinant of its competitive advantage. The goals of innovation portfolio management are (see Schentler et al. 2010, p. 310):

- Maximization of the financial value of the company's innovation portfolio aligned with the business strategy.
- Ensuring a balance among the various projects of the company in terms of risk and return.
- Fitting the number of projects to the available capacity of the company.
- Ensuring that the company's innovation portfolio matches and reflects the company's (innovation) strategy.

Performance measurement in terms of portfolio management aims to assess the target achievement in these four areas. Based on this, Cooper et al. (2001), define six key requirements for project portfolio performance:

- Projects are aligned with business' objective.
- Portfolio contains very high value projects.
- Spending reflects the business' strategy.
- Projects are done on time.
- Portfolio has a good balance of projects.
- Portfolio has the right number of projects.

Financial measures such as return on investment, net present value, or internal rate of return are suitable to evaluate the overall financial value of the company's innovation portfolio. However, they come with some drawbacks that were elaborated in Chapter "Valuing Research and Development Projects in Energy Markets (Schäfer)" of this book. The quantitative measures can be supported by qualitative ones such as peer reviews and mental checklists. Best performers in innovation management use explicit formalized tools in the selection process and apply these tools to all projects considered for adoption to the innovation portfolio (see Adams et al. 2006, p. 35; Cooper et al. 1999). Qualitative assessment of the existence, quality and application of such formal process selection processes should therefore be part of the measurement of performance at the portfolio management level. To measure and evaluate the composition of the portfolio, management should take a look at whether the portfolio is balanced in terms of quantity of short- and long term-projects. Also, the balance between high and low risk projects and large and small projects should be closely examined.

Important for the future success of a company is also to have a project pipeline that consists of projects in different stages. Future gaps need to be identified in order to be able to take action as soon as possible. A graphical illustration of the project portfolio pipeline is depicted in Fig. 7. The phases of the innovation process are

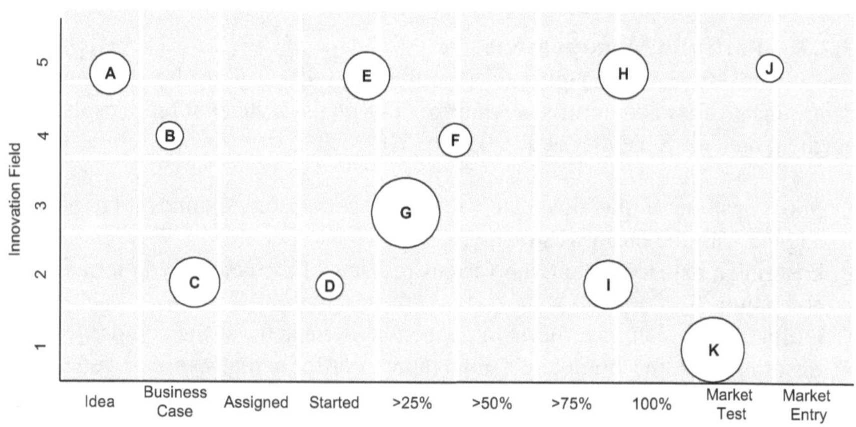

Fig. 7 Example innovation pipeline (Source: Schentler et al. 2010, p. 312)

shown on the horizontal axis. The different innovation fields, e.g., the sectors of the organization are visualized on the vertical axes. The planned turnover is represented by the size of the circle.

3.3 Putting the Framework into Action

We have described seven categories that influence innovation performance and therefore need to be considered in an innovation performance measurement framework: these categories focus on the one hand on the specific components of the innovation process, for example, input, process, output, and outcome. On the other hand, they account for contextual factors such as strategy, organization and culture, knowledge management and portfolio management. Thus, the framework ensures a holistic view of innovation and allows the capture of all the different levers that drive innovation power of a company. The design of the framework conveys the notion that innovation performance cannot be expressed via quantitative metrics only. Qualitative indicators are indispensable for describing the contextual situation that influences innovation power and should therefore be part of any innovation performance measurement system—especially when assessing innovation power at the firm-level. The uniqueness of innovation activities does not allow a one-size-fits-all approach for measuring innovation performance. The presented framework can be regarded as a general structure that gives orientation when assessing innovation power. However, it always needs to be adapted to the specific context and background of the company: external factors like market environment, competitors and the overall economic situation heavily influence commercialization and the outcome of the innovation process. Internal factors such as company size, technologies and innovation strategy influence the innovation process. Both internal and external factors influence the design of the framework and the selection of the metrics. The definition of how information is gathered and the assessment is conducted also plays an important part. Qualitative data can, inter alia, be collected via self-assessments, expert interviews, and benchmarks.

Having evaluated innovation performance, it is key to make use of the collected metrics and data. One of the objectives of performance measurement is to provide past- and future-oriented management control information to support continuous improvement. The process of performance measurement therefore needs to be followed by in-depth analysis to identify areas of weaknesses and strength. This information can then be used for supporting strategic decision-making of top management, e.g., decisions on resource allocation and future technology roadmaps. Only then can the total potential of innovation performance measurement be developed to the full.

In this section, we have focused on the theoretical aspects on innovation performance measurement albeit with a close look at practical implementation. The next section will further reinforce the practice-related focus of this book by presenting a case study that describes the process of assessing innovation power at Siemens. The case study underlines the applicability of the described performance measurement framework and discusses the implementation of theoretical findings into corporate practice. We will first point out the standing of innovation at Siemens to create an understanding of the contextual factors, followed by a detailed discussion of the Siemens innovation review process.

4 Case Study: Evaluating Innovation Power at Siemens

4.1 Role of Innovation at Siemens

Innovation has a high standing at Siemens. Innovation and engineering excellence are among the core values that Siemens stands for and secure the technological basis of Siemens. It is innovation that ensures that Siemens stays successful. Strength in innovation power is indispensable for Siemens to reach the ambitious long-term goals set in the corporate strategy. One of the three strategic directions of Siemens is to focus on innovation-driven growth markets. In order to succeed in these markets, the innovation strategy of Siemens is to be a pioneer in all of its businesses. This means being ahead of competitors in pace-setting technologies, key technologies as well as basic technologies. Siemens' innovation activities are geared towards ensuring economically sustainable energy supplies and developing software solutions. These are essential to maintaining Siemens' long-term competitiveness.

Two innovative product developments that underline these objectives have already been discussed in this book: the development of highly efficient gas turbines in Chapter "Valuing Research and Development Projects in Energy Markets (Schäfer)", and the development of software-based solutions for product life-cycle management and digital product development, discussed in Chapter "Valuing Product Lifecycle Management (Scheubel, Bierschneider, Gierse, Hermann, Wokusch)". Improving low-loss electricity transmission systems, developing new solutions for smart grids, making medical imaging and healthcare IT an integral part of outcome-oriented treatment plans are only some of the different focus areas of Siemens innovation activities. To ensure innovativeness, Siemens aims to have a competitive R&D intensity. In the fiscal year 2015, R&D expenses at Siemens accumulated to EUR 4.5 billion, which corresponds to an R&D intensity of 5.9%.[1] The general perception of Siemens' innovation power is at a high level. For example, Siemens has been ranked number one in the category "Innovation Management" of the Dow Jones Sustainability Index since 2012. Also in patent applications and patent grants, Siemens is among the top performers worldwide, as depicted in Fig. 8.

However, innovation is no sure-fire success. It means continuous effort and improvement of innovation power in order to prosper in a market environment that is highly competitive and characterized by global megatrends as well as an unstable macro-economic situation. Therefore, Siemens focuses on three areas of emphasis in its innovation activities:

[1] R&D intensity is defined as the ratio of R&D expenses and revenue.

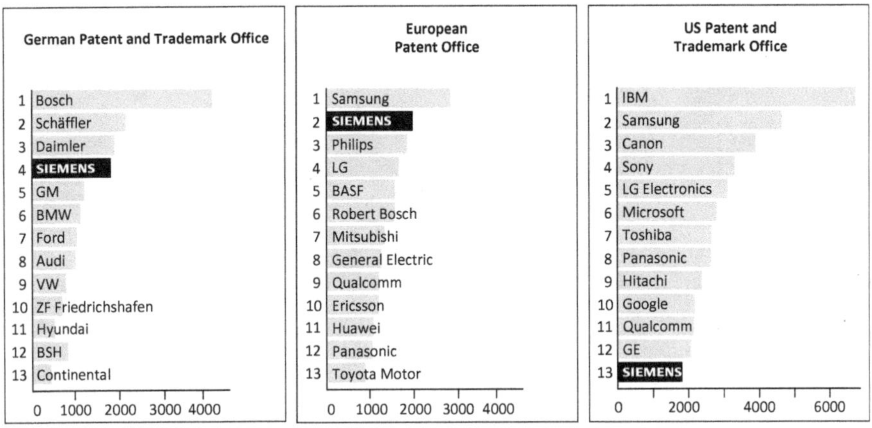

Fig. 8 Siemens in patent rankings 2013 (Source: German Patent and Trade Mark Office DPMA, European Patent Office EPO, International Intellectual Property Office IPO)

1. Ensuring long-term future viability.
2. Enhancing technological competitiveness.
3. Optimizing the allocation of R&D resources.

A clear focus and innovation strategy is one part of innovation success. The second one is to steadily control and manage innovation performance in order to secure long-term success of the company. The previous section outlined theoretical principles for measuring innovation performance at the firm level. This chapter provides a practice-related application of the discussed basics by describing how Siemens controls and measures innovation power of its businesses at the corporate level.

4.2 Assessment of the Siemens Innovation Review

4.2.1 Classification of the Innovation Review

Siemens uses a variety of tools and methods for measuring and analyzing internal innovation and R&D performance. Figure 9 gives an overview of these tools and methods for different measurement levels. Innovation performance is measured on four different levels: product line and project level, business unit level, business sector level, on the level of single divisions and on corporate or sector level. Thus, innovation performance measurement at Siemens covers all relevant levels discussed in Sect. 2.3. Depending on the measurement level, different tools and methods are used that put the focus on either efficiency or effectiveness. On product line and project level, performance is, inter alia, assessed via capability maturity model integration assessment (CMMI assessments). CMMI assessments combine proven practices and reference models and aim to support continuous process

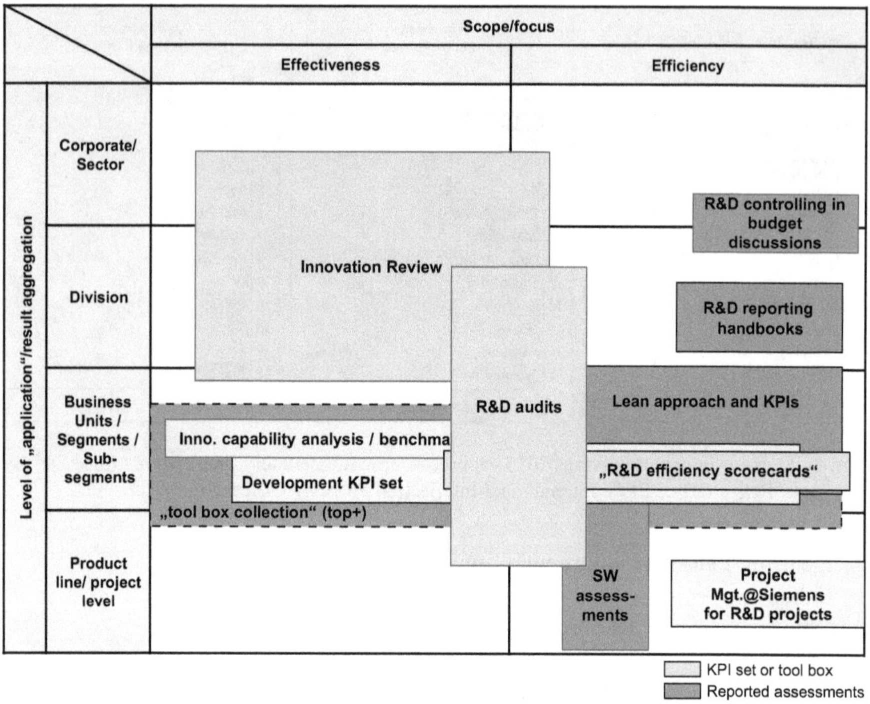

Fig. 9 Classification of innovation review in R&D assessment at Siemens (Source: Siemens AG, Technology and Innovation Strategy)

improvements. Hence, the focus of this assessment tool is primarily put on process efficiency. At the business unit level, a toolbox collection exists to measure efficiency as well as effectiveness of innovation activities. Tools include innovation capability analysis and a set of different KPIs for evaluating and supporting lean development. At the division, corporate, and sector level, innovation effectiveness is at the center of performance measurement (compare Sect. 2.3). The measurement focus shifts from operational excellence at the project level to innovation excellence. The predominant method to measure innovation performance and power at the corporate and sector level at Siemens is the innovation review. The framework of the innovation review is explained and discussed in more detail in the subsequent sections.

4.2.2 Basic Principles and Goals of the Innovation Review

The paramount purpose of the innovation review is to evaluate the innovation power of the Siemens businesses. As described above, investment in innovation power influences future profits and growth. In the high-competitive and high-technology markets that Siemens is active in, innovation power is decisive for creating products and services that meet customer's demands and outperform competitors.

For a trendsetter in new technologies and markets, a competitive portfolio of innovative products is as important as being able to rapidly recognize weaknesses in the innovation power. Assessing the innovation power of the company is, therefore, of greatest importance for Siemens. The innovation review plays a key role in this task. Its outcome gives a rationale of the development of future profit and growth of the respective business units and is, therefore, an essential part of strategic planning at Siemens.

The assessment is conducted in the form of a self-assessment on a mandatory set of criteria. This set of criteria is represented by different key questions covering the three major areas that influence innovation power at Siemens: technological competitiveness, resource allocation, and long-term orientation. On a regular basis, in general every year, responsible employees in the business units do the self-assessment.

It is important to emphasize that the assessment and the individual questions are seen as a guide to improve innovation excellence. The innovation review is a pure evaluation of the innovation performance and has no performance controlling character. This means reward or bonus systems are not based on the outcomes of the innovation review. Otherwise, participants would be incentivized to whitewash the actual situation and the objectiveness of outcomes in the self-assessment would not be ensured. The character of a pure evaluation tool with no aspect of performance controlling is shown by the clear focus on a business perspective and not at the product or technology level.

Nevertheless, answers to key questions always contain subjective components. Therefore, members of the corporate technology and innovation management department screen the results of the self-assessment using plausibility checks and, if necessary, adjust the answers (= CTO view of the company).

Siemens business units differ in structure, technology and products as well as the industry and markets they engage in. These differences need to be considered when evaluating innovation performance of the businesses. This leads to conflicting goals in the set-up of the framework. On the one hand, outcomes of the innovation review should be comparable over all units. On the other hand, the different contextual situations must be considered. Siemens overcomes this balancing act by designing the key questions of the innovation review as comprehensively and concretely as possible but not applying a "one-size-fits-all" scheme. The systematic approach leads to results that are as objective and comparable as possible. At the same time, Siemens considers the strategic business context of the assessed unit by adjusting the proposed indicators that form the criteria catalogue; this ensures the flexibility and adjustability of the framework. The proposed indicators are not mandatory but rather a guideline for the businesses to generate the key questions.

The described framework of the innovation review leads to objective, practice-proven outcomes that represent and describe a realistic image of the actual innovation performance and innovation power of Siemens. Also, the outcomes give an indication of how the innovation power could be improved. However, it is important to note that the innovation review does not contain a detailed action plan of how to deal with identified weaknesses or strengths, but rather an overview of recommended management attention topics.

4.2.3 Design of the Innovation Review: Evaluation Criteria and Categories

When investing in innovation power, Siemens considers three important categories: technological competitiveness, resource allocation, and long-term orientation. Technological competitiveness is anchored in the innovation strategy of Siemens which sees the company as a trendsetter. A leading technological position in any field that the company already is operating or plans to enter is key to being able to take over and hold the position as a trendsetter. Furthermore, only with a leading technology can future revenues and growth potential be ensured. Resource allocation is important to consider when investing in innovation power. To point out only a few points in this context: only when sufficient R&D investments are ensured for all relevant markets, can the company can gain a strong position in the respective technological field. To develop a new technology and to improve an existing technology, sufficient workforce and liquidity is necessary. Of course, pay-off of innovation investments needs to be ensured to create sustainable value. Long-term orientation is the basis for any investment of Siemens. For invest in innovation power, this includes, inter alia, that the long-term roadmap shows convincing market potential and sales opportunities and considers possible disruptions. Future market developments, technologies and competitors should be analyzed and considered in this context.

When evaluating innovation power of businesses via the innovation review, exactly these three categories are assessed by answering 14 key questions, namely technological competitiveness, resource allocation and long-term orientation. The fundamental key questions describe the respective category. For each category, there are four to five key questions to be answered. As elaborated above, the key questions are as comprehensive and concrete as possible to guarantee a systematic approach and objective outcomes. To support this idea and also to provide a guideline for answering the key questions, a criteria catalogue with around 50 indicators is provided. Thus, each key question is characterized and described by around four proposed indicators. The indicators are formulated as sub-questions. These sub-questions are not mandatory to answer but serve as a proposed guideline to evaluate the key question. The selection of the sub-questions varies depending on the business type and innovation strategy. This maintains flexibility of the innovation review as well as adaptability to different contextual factors. The criteria catalogue is continually being developed further by best-practice sharing with the sectors evaluated. The following example underlines the logic of key questions and proposed indicators.

Category	Technological competitiveness
Key question	Are customers satisfied?
Sub-questions	Sufficient consideration of customers' needs and expectations?
	Overall market share and order intake compared to competitors?
	Good customer feedback—competitive net promoter score?

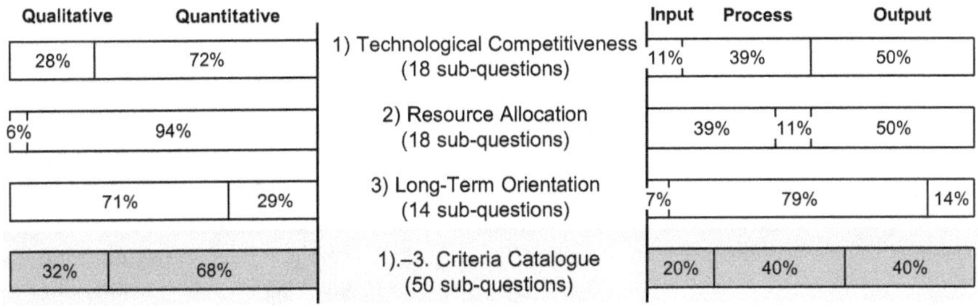

Fig. 10 Distribution of 50 sub-questions with respect to type and data format (Source: Siemens AG, Technology and Innovation Strategy)

The sub-questions refer to either the innovation input, process, or output/outcome of the respective business unit. The data format can either be qualitative or quantitative. Figure 10 shows the distribution of the 50 sub-questions with respect to type and data format. It is apparent that the number of indicators or questions referring to input, process, and output is—not perfectly, but fairly—balanced, as is the ratio between quantitative and qualitative. This underlines the holistic view of the innovation review and ensures validity of evaluation outcomes.

Innovation power in each field (i.e. the answer on the key question) is determined based on three performance levels:

1. Leading position/good practice
2. Comparable/competitive position
3. Management attention

Management attention means that the topic already is or should be tracked carefully with regard to the company's innovation power. For each key question, specific evaluation criteria and thresholds are defined. They can be of qualitative or quantitative nature and guide the assessor in choosing the appropriate performance. For each category, i.e. technological competitiveness, resource allocation, and long-term orientation, the performance levels in each key question are aggregated. This simple way of evaluating innovation power has several advantages: It is simple to understand for all participating assessors at the business unit level. Furthermore, it creates conformity over different business units. The definition of different indicators and evaluation criteria allows adaptability to different contexts and business environments of the business units. This is basic for comparability over all different business units of Siemens.

The concentration on three categories allows direct communication of the results of the innovation review to the board and top management. The limitation of the evaluation to three possible performance levels certainly results in some fuzziness. However, the layer-by-layer structure allows a quick overview of the innovation performance of the business units and analysis at the level of the 14 key questions or at the level of sub-questions. Table 1 gives an overview on the 14 key questions

Table 1 Overview on 14 key questions and proposed indicators (Source: Siemens AG, Technology and Innovation Strategy)

Assessment area	Key question	Proposed indicators
I. Technological competitiveness	Leading technology position compared to competitors?	• Technological USPs • Coverage of technology portfolio • Product performance, quality, cost
	Leading technology position compared to competitors?	• Technological USPs • Coverage of technology portfolio • Product performance, quality, cost
	All relevant markets addressed with appropriate products and technologies?	• Regional market positions (M1–M4) • SMART product pipeline • Services and IT solutions
	Customers satisfied with offers (i.e. products, services)?	• Consideration of customer needs • Market shares and order intake • Net Promoter Score (NPS)
	Current and future position protected by IPR and standardization activities?	• Strong patents in all key technology • Freedom to operate • Standardization activities
II. Resource allocation	Sufficient R&D investments ensured for all relevant markets?	• Abs./rel. R&D vs. competitors • R&D project coverage for region/ market specific products
	Efficient use of platforms, synergies and effective make-or-buy activities?	• R&D for platform development • Projects w/suppliers • Technology motivated M&A
	Sufficient and appropriate workforce available on a global basis?	• Employee turnover rate • Availability of (key) talent • Regional R&D footprint
	Strategic innovation fields driven and developed systematically?	• R&D allocated to innovation fields • Status of innovation pipeline • R&D per market growth phases
	Pay-off of innovation investments into products and services ensured?	• Revenues with new products • New product launches • Accuracy of R&D plans (t, €)
III. Long-term orientation	Long-term technological USPs and disruptions covered by roadmap?	• Endangered technological USPs • Threats by emerging competitors • Disruptive topics and white spaces
	Comprehensive information about future market developments?	• Topics in trend radar • Outside-in information • Planning scope and horizon
	Vision of future technologies, markets and competitors derived systematically?	• Innovation strategy (e.g. trendsetter) • Visioning/scenario projects • Emerging technology monitoring
	How is the innovation strategy implemented operationally?	• Strategic competence development • Innovative ability of organization • Comprehensive implementation

and proposed indicators to answer the key questions. In the subsequent section, the assessment of one key question in each of the three main categories will be explained in more detail to create an understanding of the innovation review process.

4.2.4 Technological Competitiveness

Siemens operates in highly competitive business segments. Sustainable value creation at Siemens includes continuous improvement relative to the competitors in these segments and markets. Only then can the position of a pioneer or trendsetter be achieved and fostered. This goal is, as elaborated above, integrated as a principle in the innovation strategy. A key factor for achieving innovation power for Siemens is to be technologically competitive. Assessing technological competitiveness, therefore, is one of the three categories that Siemens covers in its innovation review. In order to achieve technological competitiveness in terms of innovation power, Siemens focuses on five different factors:

1. Are we holding a competitively leading technology position?
2. Are efficient core processes established?
3. Are all relevant markets addressed with appropriate technology?
4. Are customers satisfied with offers?
5. Is our position sufficiently protected by intellectual property rights and standardization activities?

The categories show that technological competitiveness goes beyond holding a leading technology position but also requires having efficient processes, protecting the position by property rights and offering products that satisfy customers. In order to give an insight into the way how Siemens evaluates innovation power in the context of technological competitiveness, we will discuss one of the five key categories in more detail, namely, whether Siemens holds a leading technology position compared to competitors (key question 1). Several aspects like performance, costs and quality have to be optimized to achieve technological competitiveness or at least economic viability of a product or process. The objective of assessing the technology position relative to competitors is to combine all these aspects and gain insight in the technological position of products and respective manufacturing technologies. The sub-question in the criteria catalogue, which helps assessors to understand and correctly answer the key question, specifies the latter by asking for a general benchmarking of products/offerings with current top competitors with respect to performance, quality and cost. Hence, this measure can be classified as one that focuses on innovation outputs and has a qualitative data format. It is task of the business unit to define the dimensions of the required benchmark. This allows adaptability over all different processes. For example, a benchmark of industry services differs from one of healthcare products or infrastructure and network solutions. Indicative dimensions that need to be compared to the main competitor can be functionality, performance, cost, product quality, service quality, and price. These dimensions are then ranked on a scale that ranges from −2 for "very far behind competitors" to +2 for "very far ahead of competitors.

Based on this ranking, the evaluation of innovation power of the respective business unit takes place. The evaluation criteria are defined in the criteria catalogue. The innovation power can be evaluated as "leading position/good practice" if the products of the business unit show evident leadership in industry benchmarks, the business unit undertakes benchmarks frequently, and potential gaps are known, identified and documented with corrective actions. The innovation power of the business unit can be evaluated as "comparable/competitive position" if benchmark results are on par with competitors, the business unit undertakes regular benchmarks and has identified gaps. "Management attention" is the appropriate evaluation if benchmark results are behind competitors, the business unit undertakes no benchmarks or only on a random basis and only partially knows about gaps compared to competitors.

4.2.5 Resource Allocation

Providing sufficient financial resources is one of the major influences on innovation power. We briefly outlined the necessity of sound funding of innovation projects in Sect. 3.2. Funding of innovation always goes in hand with efficient allocation of resources. Particularly for Siemens, consisting of a portfolio of different businesses and being confronted with huge number of decisions regarding innovation activities, efficient resource allocation is key for sustainable value creation. Therefore, the assessment of resource allocation is part of the innovation review. For Siemens, five key categories are of importance in this context:

1. Is sufficient R&D investment ensured for all relevant global markets?
2. Do we make use of platforms, synergies and make-or-buy activities?
3. Is sufficient workforce available on a global basis?
4. Are innovation fields driven and developed systematically?
5. Do innovation investment into products and services pay off?

To specify the process of assessing resource allocation, we will discuss one of these five key question in more detail. In Sect. 2, we explained that innovation portfolio management is important to ensure a company's future revenue potential and we elaborated on the necessity of innovation portfolio management being a part of an innovation performance measurement system. Siemens accounts for this by assessing whether innovation fields are driven and developed systematically (key question 4). More concretely, we evaluate whether the business unit has sufficient R&D investment and products for all phases of the market life-cycle.

The market life-cycle is a basic model for describing the revenue potential of a product. After the development phase with no revenues, the product starts generating revenues with its introduction. In the growth phase, revenues constantly increase and reach their maximum in the phase of maturity. The last phase of the market life-cycle, the so called harvest phase, is typically characterized by decreasing revenues. At Siemens, the objective of any business unit in terms of innovation portfolio

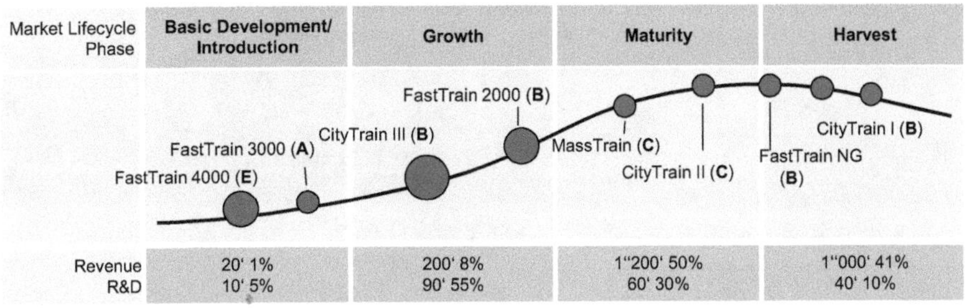

Fig. 11 Visualization of product lifecycle and innovation pipeline (Source: Siemens AG, Technology and Innovation Strategy)

management is to have the product and innovation pipeline always filled with products at the different stages. These products should outperform those of competitors. Figure 11 visualizes the evaluation principle of the innovation review for one of the Siemens division. The size of the bubble indicates the expected R&D investments (accumulated over the next 5 years), and the letter represents the expected accumulated revenues over the future 5 years (A = high revenues to E = low revenues).

The evaluation criteria for the innovation power of Siemens Rail Systems in terms of resource allocation along the market life-cycle are defined as follows. Innovation power is "leading position/good practice" when the business unit is characterized by a growth focus and has sufficient investment in new products as well as a sufficient number of product lines and projects in all market phases. Expected investment and outcome should be superior to competitors. If the distribution is largely balanced with only minor gaps for some market phases, and investment and expected outcome is comparable to competitors, the innovation power of the business is categorized as "comparable/competitive position". "Management attention" is necessary if the focus of the business unit is on the harvest phases and significant gaps in the early market phases do exist. Investment and expected outcomes that are inadequate compared to competitors are also signs of innovation power that requires management attention.

4.2.6 Long-Term Orientation

In organizations with a short-term orientation, investment in innovation power underlies a trade-off between profit and growth since the degree of R&D intensity is a direct lever on the company's profit margin. A short-term orientation therefore favors low R&D investment and tilts the innovation portfolio to short-term projects that tend to be incremental and safe.

However, strong innovation performance secures a company's future growth and results in increasing long-term profits. Figure 12 graphically illustrates the short-term trade-off between investment in innovation power, growth and profit.

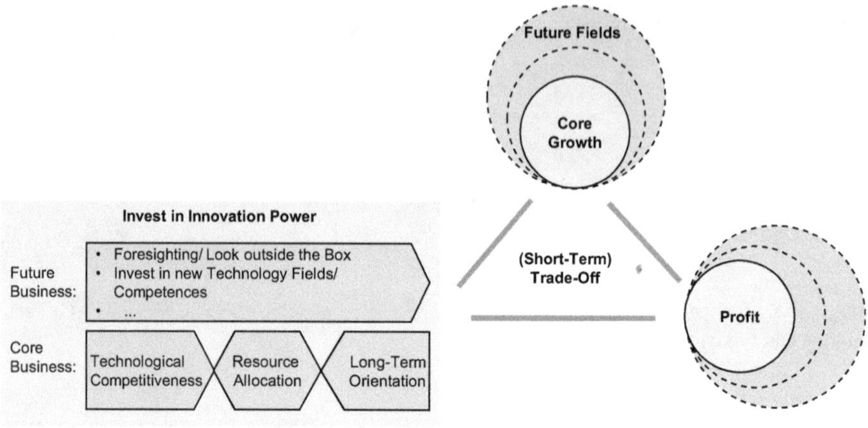

Fig. 12 Trade-off between investment in innovation power, profit and growth (Source: Siemens AG, Technology and Innovation Strategy)

To remove the risk of exchanging future growth opportunities in favor of short-term profit increase, Siemens consistently follows a long-term innovation strategy that is consistent with the Siemens business strategy. Long-term orientation of innovation activities is therefore one of the three categories in the Siemens innovation review. Siemens has identified four key questions that are important for assessing its innovation power:

1. Does the roadmap cover long-term technological unique selling propositions and disruption?
2. Is there comprehensive information about future market development?
3. Is there a clear vision about future technology, markets, and competitors?
4. How is the innovation strategy implemented operationally?

In the following, we will focus on key question four and briefly describe the Siemens approach to evaluating how its innovation strategy is implemented operationally. As already elaborated, business and innovation strategies need to be consistent. For Siemens, taking a holistic point of view implies to going one step further and not only aligning innovation and business strategy but all types of roadmaps and respective projects in order to both ensure and optimize the overall success of the strategy. In terms of evaluating the implementation of the innovation strategy, Siemens therefore assesses the alignment of the five most important topics of its strategic roadmap, technology roadmap, innovation roadmap, competence development roadmap and strategic measures. The goals of the innovation strategy can only be realized if the priorities of topics within the different roadmaps and the defined measures for anticipated threats and moves by competitors are aligned, which results in the alignment of the work activities of all parties.

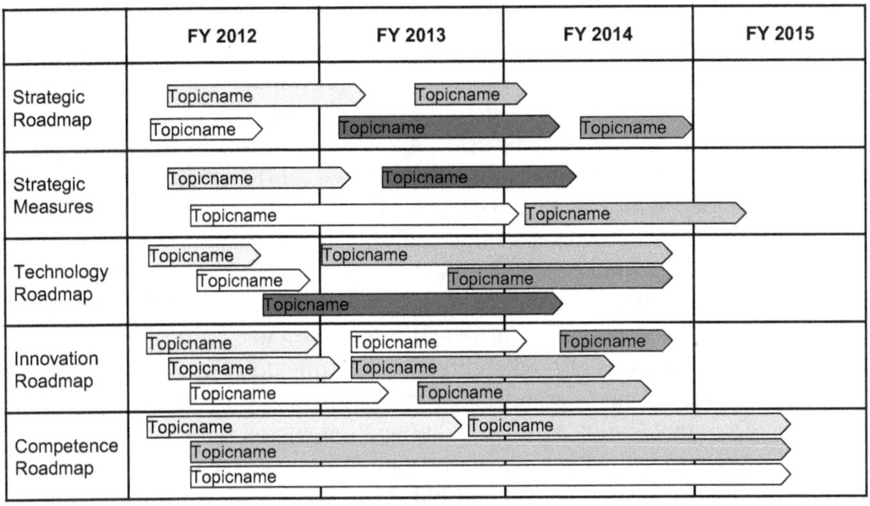

Fig. 13 Alignment of various roadmaps (Source: Siemens AG, Technology and Innovation Strategy)

To evaluate the extent to which this goal is achieved, the business units prepare a synopsis of the roadmap topics for the technology, innovation and strategic roadmaps as well as associated strategic measures and denote the interrelation between the topics. This is graphically depicted in Fig. 13. The grade of interrelation is illustrated by color coding.

Based on this synopsis, the consistency of the various roadmaps is assessed. A business unit that is evaluated as "good practice" typically shows alignment of the strategic and innovation roadmap. Also, sufficient measures are defined to anticipate the major market and competitor moves. "Management attention" is necessary if there are contradictions between various roadmaps and measures are largely reactive or not consistent.

The described evaluation procedure is a comprehensible way of evaluating the implementation of the innovation strategy. The focus on the five most important topics of the respective roadmaps not only prevents assessors from getting lost in detail, but also shifts attention to those points that influence the success of the strategy implementation the most.

4.2.7 Analysis of Innovation Review Outcomes

In the previous sections, we have described the concept and process of the Siemens innovation review in detail. We have explained that the innovation power of any business unit is described by its performance in the three key categories technological competitiveness, resource allocation and long-term orientation. By clustering the evaluation outcomes into three different performance levels, the outcomes directly give an indication of weaknesses and strengths. However, no detailed action plan is defined in the self-assessment process. Rather, as is the purpose of the performance measurement system described in Sect. 3, the innovation review provides past- and future-oriented management control information and provides strategic insights to better support continuous improvement and decision-making. In the following, we will briefly describe how the outcomes of the innovation review are used to support strategic decision-making at Siemens. In order to derive actions based on the results of the innovation review, a differentiated analysis of the business units depending on their specific business situation is necessary.

Two groups are focused on in this analysis: business units with high profit margins and business units with low profit margins. Business units with high profit margins are today's profit engines, i.e. they significantly add to the profitability and value creation of Siemens. Business units with low profit margins are underperforming in terms of profitability, but can still add a significant proportion to overall revenues or have a significant portfolio impact by, for example, technical synergies. The combination of profitability and innovation power of the business units is illustrated to provide top management with a meaningful and convincing summary of the innovation review (see Fig. 14).

The horizontal axis represents the revenue of the business unit, the vertical axis the profit. The business units are sorted according their profit margin in descending order. The outcomes of the innovation review are allotted to each business unit. Additionally, for each business unit the relevance of technology for business success is determined as either low, medium or high. Based on this combination of profit margin and innovation power, different types of management attention and further actions for each business unit can be derived. Specific attention is paid business units which are evaluated with "management attention" in at least one of

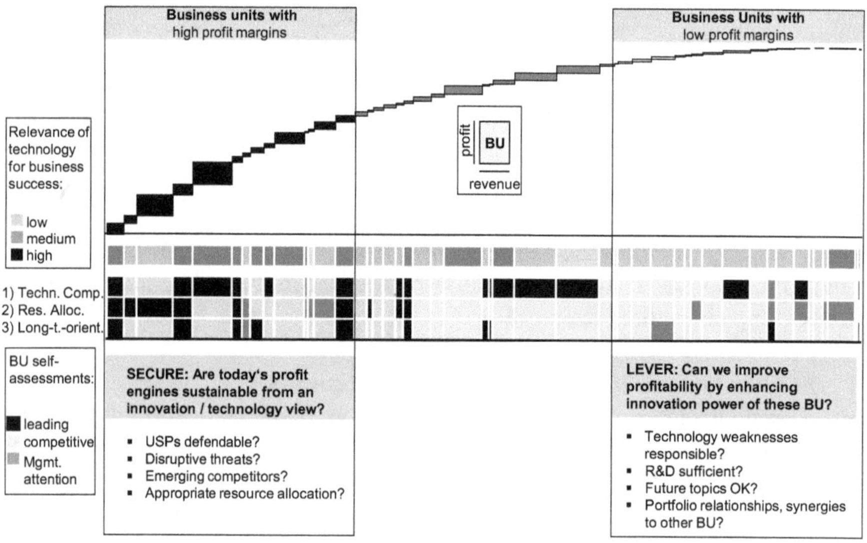

Fig. 14 Financial contributions of business units sorted by profit margin and innovation power assessments (Source: Siemens AG, Technology and Innovation Strategy)

the three assessed categories. The key question is whether these profit engines are sustainable from an innovation and technology perspective. If a business unit lacks in innovation power, this can jeopardize its future success and profitability. High profit margin businesses then can easily become low profit businesses. This has consequences on the overall Siemens portfolio, for example, consequences on overall profitability, revenues, risk and market position.

When considering a long-term (innovation) strategy, critical assessment of the categories with poor innovation power is necessary. In this context, impacts of poor innovation power of the respective business units should be assessed and further actions defined. The goal of this in-depth assessment and action definition is to secure the business units' sustainability from an innovation and technology point of view. Siemens applies four different questions which help to support this process.

1. Are technological unique selling propositions (USP) the basis for business success, and can we secure them in the future?

   If business units need to defend their USPs or see those replaced by competitors, Siemens checks the strategic position and R&D program of these business units. The technology and product roadmap is also reviewed and compared to those of competitors.
2. Are there mid- and long-term threats arising from trends or disruptions?

   In this case, specific attention is paid to long-term risk management. Anticipation of these disruptive changes and screening of the emerging technology as well as the own technology is part of the action plan. To react to threats and disruptions, a counter strategy is defined.
3. Do new, emerging, or aggressive competitors challenge us with their technology and innovation power?

   In case of risks coming from aggressive competitors, those are analyzed in detail. If relevant, the own technology is strengthened in order to differentiate from or outperform the competitor.
4. Is the resource allocation adequate to the Siemens innovation strategy?

   It can be necessary to optimize the R&D allocation in order to ensure sustainability. Appropriate levels of R&D spending should be ensured to secure the technical USPs or prepare for disruptions. Risks from cost pressure often have consequences on R&D investment. This must be considered in R&D allocation.

4.2.8 Recommended Management Attention for Business Units with Low Profit Margins

Business units with low profit margins do not allocate capital and resources of shareholders in the most efficient way possible. In terms of the innovation review, business units with low profit margins are therefore analyzed in more detail. Low profit margins do not necessarily go in hand with poor innovation power. It might simply be a characteristic of the market segment. However, strengthening innovation power can have an effect on profitability. There are many ways, for example, by increasing R&D or production efficiency, gaining market share through advanced technologies which go in hand with scale effects, or by entering higher price segments with products that outperform competitors. The purpose of an in-depth assessment of business units with low profit margins and poor innovation power (= evaluated as "management attention), hence, is to analyze whether profitability of these business units can be improved by enhancing their innovation power. Similarly to the procedure for high margin business units, four key questions serve as a guidance for the detailed assessment of these business units.

1. Are technological weaknesses the reason for low profitability? Do we have the right products, processes and costs?

   Own weaknesses in technological and innovation power in comparison to the best competitors need to be assessed. Checking the roadmap of the business unit and identifying technological USPs is a possible way to analyze the position of the business unit. Realization of these potentials is the subsequent action that follows the analysis.
2. Is R&D sufficient compared to competitors or low due to margin pressure?

   Business units with a critical tradeoff between invest in R&D and profitability goals should align R&D spending to the long-term business goals and competitors spending. Start-ups and fast growing businesses should be considered separately since contextual factors influence specifications.
3. Which future topics are planned and are they funded adequately with R&D?

   Promising projects can help to overcome a lack in profitability in the near future. For Siemens businesses with a critical pipeline of future topics, roadmaps and allocation of R&D budgets for development of future offerings are challenged. Critical R&D projects do receive special observation and are fixed, for example, strengthened or cancelled.
4. Are there portfolio relationships to other business units?

   If business units are seen to have difficulties in building up a sustainable USP from a technological and innovation perspective, portfolio impacts to other business units are analyzed in more detail, for example, the relevance of the specific business unit's products for other Siemens divisions and technical synergies. The analysis of portfolio relationship is the basis for being able to take action, for example, restructuring or divestment.

4.3 Summary and Discussion

The success of Siemens is dependent on the innovation performance and innovation power of its business units. In order to be able to take the role of a pioneer in innovation-driven growth markets, Siemens must be ahead of competitors in terms of technology and process efficiency. Strong innovation power is one of the most important preconditions for reaching this ambitious goal and to ensure long-term success and sustainable value creation for the organization. For Siemens, it is a continuous challenge to sustain and foster innovation activities.

As discussed in Sect. 3.2, there are a number of important factors for ensuring innovation power of a company. Siemens has identified three main categories that heavily influence innovation outcomes. First, technological competitiveness. Good performance in this key area is a basic requirement for survival against competitors. However, this is not enough to ensure technological competitiveness. The first additional factor needed is the establishment of efficient core processes to efficiently and effectively develop ideas into new products. Secondly, different markets should be addressed with the appropriate technology, since customers in emerging markets can have significantly different needs than those in industrialized countries. Thirdly, customer satisfaction goes in hand with appropriate technology, and is therefore also an indicator for technological competitiveness. Finally, the technology must be protected against competitors by intellectual property rights and standardization processes. These factors all form a part of technological competitiveness.

The second category that Siemens has identified as being of major importance for innovation power is resource allocation. This goes beyond sufficient funding of R&D projects in all relevant markets but also includes ensuring an appropriate pay-off of innovation investments. Efficiently allocating the available resources is key for maximizing the overall company value. Therefore, the strategic innovation fields in particular need to be driven and developed systematically. Using platforms and synergies within the group as well as having effective make-or-buy activities drives efficiency of innovation processes and hence are an integral part of resource allocation at Siemens. Furthermore, resource allocation must ensure that sufficient and appropriate workforce is available on a global basis; a regional R&D footprint can positively support this goal.

The third category is long-term orientation. The short-term tradeoff between investment in innovation power and long term growth on the one hand and short term profit on the other hand can only be overcome by aligning the business, innovation, and technological strategies to long-term goals and ensuring consistency of the different roadmaps that influence innovation activities. In this context, future market developments and actions of competitors need to be anticipated and derived systematically.

The innovation review of Siemens aims to assess the performance of business units in these three categories, i.e. technological competition, resource allocation, and long-term orientation. The innovation review is conceived as a self-assessment of the different business units. Experts of the corporate technology department serve as control instances and ensure objectivity of the outcomes. Objectivity is further secured by the systematic approach of the review with key questions formulated as concretely and comprehensively as possible. The outcomes give a rationale for the future development of profit and growth and help to identify further actions.

The innovation review at Siemens takes the different contextual factors of its business units into account. This is done by considering the strategic business context of the assessed unit in the application of the proposed indicators. The application of the indicators is not mandatory but rather serves as a guideline. In order to generate outcomes that are comprehensible and communicable, the evaluation is limited to three performance levels: "leading position/good practice", "comparable/competitive position" and "management attention". We have described that, for Siemens, two groups of business units are of specific importance in an in-depth analysis of the evaluation outcomes. Namely, business units with high profit margins and low profit margins that are evaluated with "management attention" in one of the three main categories. We need to assess whether today's profit engines are sustainable from an innovation and technology point of view and whether profitability of low margin businesses can be improved by enhancing their innovation power.

In Sect. 3 we have presented the basic challenges and requirements for measuring and evaluating innovation performance at the firm level. We have described our measurement framework which focusses on process-related factors such as input, output and outcome as well as on contextual factors like business strategy, innovation culture, and knowledge management. This holistic view and the combination of quantitative and qualitative assessment criteria were found to be key requirements for measuring innovation performance.

The detailed explanation of the Siemens case study showed that these empirically and theoretically grounded requirements seem to be applied and implemented in corporate innovation management. Surveys show that there is consensus among executives that innovation should be measured rigorously but less than half of the companies do so. Our theoretical discussion of the basic requirements for measuring innovation performance in combination with the practice-related excursions to innovation management practice at Siemens can help to overcome this obstacle. The explanations can serve as a motivation for measuring innovation performance and serve as a guideline for the effective setup of a framework to assess innovation power.

References

Abbey, A., & Dickson, J. (1983). R&D work climate and innovation in semi-conductors. *Academy of Management Journal, 26*, 362–368.

Adams, R., Bessant, J., & Phelps, R. (2006). Innovation management measurement: A review. *International Journal of Management Reviews, 8*(1), 21–47.

Amabile, T. M., Conti, R., Coon, H., Lazenby, J., & Herron, M. (1996). Assessing the work environment for creativity. *Academy of Management Journal, 39*, 1154–1184.

Andrew, J. P., Haanæs, K., Michael, D. C., Sirkin, H. L., & Taylor, A. (2009). *Measuring innovation 2009: The need for action. BCG report*. Boston: The Boston Consulting Group Inc.

Arvanitis, S., Lokshin, B., Mohnen, P., & Woerter, M. (2015). Impact of external knowledge acquisition strategies on innovation: A comparative study based on Dutch and Swiss Panel data. *Review of Industrial Organization, 46*, 359–382.

Balachandra, R., & Raelin, J. A. (1980). How to decide when to abandon a project. *The International Journal of Research Management, 23*, 24–29.

Bleicher, F. (1990). *Effiziente Forschung und Entwicklung: Personelle, organisatorische und führungstechnische Instrumente*. Wiesbaden.

Bougrain, F., & Haudeville, B. (2002). Innovation, collaboration and SMEs internal research capacities. *Research Policy, 31*(5), 735–747.

Brown, M. D., & Svenson, R. A. (1998). Measuring R&D productivity. *Research Technology Management, 41*(6), 30–35.

Chiesa, V., Frattini, F., Lazzarotti, V., & Manzini, R. (2009). Performance measurement in R&D: Exploring the interplay between measurement objectives, dimensions of performance and contextual factors. *R&D Management, 39*(5), 487–519.
Cooper, R. G., Edgett, S. J., & Kleinschmidt, E. J. (1999). New product portfolio managment: Practices and performance. *Journal of Product Innovation Management, 16*, 333–351.
Cooper, R. G., Scott, J. E., & Kleinschmidt, E. J. (2001). Portfolio management for new product development. *R&D Management, 31*(4), 361–380.
Damanpour, F. (1991). Organizational innovation: A meta-analysis of effects of determinants and moderators. *Academy of Management Journal, 34*, 555–590.
Damanpour, F. (1996). Organizational complexity and innovation: Developing and testing multiple contingency models. *Management Science, 42*, 693–716 (1991).
De Leede, J., Nijhof, A. H., & Fischer, O. A. (1999). The myth of self-managing teams: A reflection on the allocation of responsibilities between individuals, teams and the organization. *Journal of Business Ethics, 21*, 203–215.
de Spiegelaere, S., van Gyes, G., van Thillo, T., & van Hootegem, G. (2012). *Innovative work behavior: Concept and measurement*. Proceedings of the XXIII. ISPIM conference: Action for Innovation, Innovating from Experience, 2012.
Dodgson, M., & Hinze, S. (2000). Indicators used to measure the innovation process: Defects and possible remedies. *Research Evaluation, 8*, 101–114.
Ekvall, G. (1996). Organizational climate for creativity and innovation. *European Journal of Work and Organizational Psychology, 5*, 105–123.
Gamal, D., Salah, T., & Elrayes, N. (2011). *How to measure organization innovativeness?* (Working Paper). Technology Innovation and Entrepreneurship Center Egypt.
Gleich, R. (2011). *Performance Measurement – Konzepte, Fallstudien und Grundschema für die Praxis*. München: Vahlen.
Globe, S., Levy, G. W., & Schwartz, C. M. (1973). Key factors and events in the innovation process. *Research Management, 16*, 8–15.
Grenzmann, C., Marquardt, R., Revermann, C., & Wudtke, J. (2004). *Forschung und Entwicklung in der Wirtschaft – Bericht über die FuE-Erhebungen 2001 und 2002*. Essen.
Hansen, M. T., & Birkinshaw, J. (2007). The innovation value chain. *Harvard Business Review, 85* (6), 121–130.
Hauschildt, J. (1992). Ist das Rechnungswesen innovationsfeindlich? In K. Boysen, G. Hohlfeldt, H.-J. Jacob, F. Nehles, & R. Wellman (Eds.), *Der Wirtschaftsprüfer vor innovativen Herausforderungen – Festschrift für Hans-Heinrich Otte* (pp. 52–67). Stuttgart: Schäffer-Poeschel.
Janssen, S., & Möller, K. (2011). Erfolgreiche Steuerung von Innovationsprozessen und –projekten – Ergebnisse einer emprischen Studie. *ZFCM, 55*(2), 97–104.
Kerssens-van Drongelen, I. C., & Cook, A. (1997). Design principles for the development of measurement systems for research and development. *R&D Management, 27*(4), 345–357.
Kerssens-van Drongelen, I. C., & Cook, A. (2000). Performance measurement in industrial R&D. *International Journal of Management Reviews, 2*(2), 111–143.
Kuratko, D. F., Hornsby, J. S., & Hayton, J. (2015). Corporate entrepreneurship: The innovative challenge for a new global economic reality. *Small Business Economics, 45*(2), 254–253.
Lee, M., Son, B., & Lee, I. (1996). Measuring R&D effectiveness in Korean companies. *Research Technology Management, 39*(1996), 112–132.
Littkemann, J. (2005). *Innovationscontrolling*. München: Vahlen.
Möller, K., & Schönefeld, C. (2011). Innovation performance measurement – ein Referenzmodell zur Analyse der Innovationssteuerung. *Controlling & Management, 55*, 367–371.
Morris, L. (2008). *Innovation metrics – The innovation process and how to measure it* (Working Paper). Innovation Labs LLC.
O'Reilly, B., & Rao, R. M. (1997). The secrets of America's most admired corporations: New idea, new products. *Fortune, 135*, 60–64.

Ochoa, R. L., & Peña, J. I. (2012). *Measuring innovation: Trends and characteristics*. Proceedings of the XXIII. ISPIM Conference: Action for Innovation, Innovation from Experience 2012.

Schentler, P., Lindner, F., & Gleich, R. (2010). Innovation performance measurement. In Gerybadze et al. (Eds.), *Innovation and international corporate growth*. Heidelberg: Springer.

Schumpeter, J. A. (1934). *The theory of economic development*. Cambridge: Harvard University Press.

Stippel, N. (1999). *Innovations-controlling*. München: Vahlen.

Stock, G. N., Greis, N. P., & Fischer, W. A. (2001). Absorptive capacity and new product development. *Journal of High Technology Management Research, 12*, 77–91.

Stone, A., Rose, S., Lal, B., & Shipp, S. (2008). *Measuring innovation and intangibles: A business perspective*. Washington, DC: Institute for Defense Analyses (IDA).

Sundbo, J. (1997). Management of innovation in services. *Service Industries Journal, 17*, 32–455.

Trautmann, K., & Enkel, E. (2014). Success factors for strategic communication of corporate innovativeness for financial analysts. *International Journal of Innovation Management, 18*(1), 1–46.

Utterback, J. (1971). The Process of innovation: A study of the origination and development of ideas for new scientific instruments. *IEEE-Transactions on Engineering Management, 18*, 124–131.

Werner, B. M. (2002). *Messung und Bewertung der Leistung von Forschung und Entwicklung im Innovationsprozess*. Dissertation, Darmstadt.

Werner, B. M., & Souder, W. E. (1997). Measuring R&D performance state of the art. *Research Technology Management, 40*(3), 28–32.

Zahra, S. A., & George, G. (2002). Absorptive capacity: A review, reconceptualization, and extension. *Academy of Management Review, 27*, 185 ff.

Conclusions

Gunther Friedl and Horst J. Kayser

Valuing corporate innovation is a broad field of investigation. This work has focused on three important and current topics to provide findings about the state of the art methods and tools of valuing innovation.

It started with concepts for valuing new product development projects in the presence of uncertainty and managerial flexibility. Academic research suggests the usage of decision tree analysis and real option models. They are more appropriate for incorporating different types of risks and the flexibility to react to certain developments. While the case study demonstrates that Siemens does not directly apply these concepts according to the textbooks, the valuation process at Siemens fully and successfully incorporates a detailed risk assessment and the staging of the entire development project. Thus, the valuation process is close to the suggestions from academia, though more pragmatic in the implementation.

The second field of interest was product lifecycle management. Digitalization affects all elements of the value chain, including research and development. Product lifecycle management has the potential to increase the creative power of engineers, accelerate innovation processes, and reduce the cost of new product development. While the theoretical considerations demonstrate that product lifecycle management changes innovation processes in the long run, the case study at Siemens shows that implementing product lifecycle management is a large change process. But ultimately, there is no alternative to product lifecycle management if companies want to stay competitive on a global scale.

The third and probably broadest topic that was addressed in this book is how to measure the innovation power of a large organization. Theory suggests using a

G. Friedl (✉)
TUM School of Management, TU München, München, Germany
e-mail: gunther.friedl@tum.de

H.J. Kayser
Siemens AG, München, Germany

G. Friedl, H.J. Kayser (eds.), *Valuing Corporate Innovation*, Management for Professionals, DOI 10.1007/978-3-319-64864-4_5

multidimensional approach including, for example, quantitative and qualitative measures, input-, process-, and output-measures, and benchmarking results. The case study demonstrates that Siemens uses a sophisticated approach with many different instruments to address the many dimensions of innovation. This approach allows Siemens to identify the most innovative parts of the entire company, and to develop strategies based on this performance information. By doing that, Siemens is constantly able to improve the innovation power of its different activities.

These three topics illustrate state-of-the art concepts from two different point of views. Practitioners might be inspired by academic concepts with innovative ideas for valuing and controlling corporate innovation. Academics might be motivated to continue their research efforts and encouraged to transfer their new insights to corporate practice.

Index

G. Friedl, H.J. Kayser (eds.), *Valuing Corporate Innovation*, Management for Professionals, DOI 10.1007/978-3-319-64864-4

FSC
www.fsc.org
MIX
Papier | Fördert gute Waldnutzung
FSC® C083411

Kolibri 360 GmbH
Ferdinand-Jühlke-Straße 7
99095 Erfurt, Deutschland
produktsicherheit@kolibri360.de